Hired Hands

D1092557

Hired Hands:

Labour and the Development of
Prairie Agriculture, 1880-1930

Cecilia Danysk

Canadian Cataloguing in Publication Data
Danysk, Cecilia
   Hired hands: labour and the development of prairie agriculture,
1880-1930

(The Canadian social history series)
Includes bibliographical references and index.
ISBN 0-7710-2552-1

1. Agricultural labourers – Prairie Provinces – History.   2. Agriculture –
Prairie Provinces – History.   I. Title   II. Series.

HD1790.P7D35 1995   331.7'63'0971209   C95-931228-5

Typesetting by M&S, Toronto
Printed and bound in Canada

This book has been published with the help of a grant from the Social Science Federation of Canada, using funds provided by the Social Sciences and Humanities Research Council of Canada.

McClelland & Stewart Inc.
The Canadian Publishers
481 University Avenue
Toronto, Ontario
M5G 2E9

1 2 3 4 5   99 98 97 96 95

Contents

For David and Michael

Preface

This is a study of hired hands. Thousands of men worked and lived as paid labourers on the farms of others across the prairie West, assisting in the day-to-day and season-to-season round of activity that constituted the work of running a farm. Employed in the region's major industry, agricultural labourers formed the largest group of wage workers in Manitoba, Saskatchewan, and Alberta. Between 1878, when the first commercial shipment of prairie grain went to international markets, and 1929, when the collapse of grain prices signalled the end of the wheat boom, agriculture emerged and matured as the mainstay of the prairie West, shaping not only the economy but the society as well.

Farm workers were essential to the development of the western agricultural economy, and they left their mark on the evolving rural community. Yet they have become forgotten men in history, relegated to the periphery of labour and agricultural studies. Their part in the formation of the economy and society has been largely overlooked and their role in the development of the agricultural industry has been given a single dimension. This study does not intend to resurrect hired hands as neglected heroes of the rural prairies. It does, however, seek to place them in their historical context by casting light on the conditions of their life and labour and by situating them within the spectrum of the Canadian labour experience.

In researching and writing this book, I have incurred many scholarly and personal debts. John Herd Thompson directed the dissertation that laid the groundwork for this book, Louise Dechêne took on the task of co-supervisor when John moved from McGill to Duke University, and Mrs. Rubie Napier many times made my rough road smooth. Colleagues and friends across the country offered encouragement and helped me rethink and refine many aspects of the argument. In particular I thank Dianne George, Richard Basciano, Stanley Frost, Susan Button, Jane Cameron, Paul Voisey, Rod Macleod, Doug Owram, John Foster, Ann McDougall, Doug McCalla, Linda Williams, John Wadland, Joan Sangster, and Ruth Bleasdale. My mother and sister, Mabel Danysk and Louise McGrath, provided research assistance and constant moral support.

Especially heartfelt thanks go to historian and fellow traveller Bob Beal, who read the entire study many times, engaged me in rigorous debate, and provided penetrating criticism. He also compiled the data for the tables and graphs, steered me toward better writing, and even dreamed up the title. Above all, his confidence in the significance of this work sustained me.

I thank the Social Sciences and Humanities Research Council and the Max Bell Foundation for their financial assistance, and the Fonds pour la formation de chercheurs et l'aide et la soutien à la recherche and the J.W. McConnell Foundation for their scholastic support.

My greatest debt is to my sons, David and Michael Schwieger. Without their unwavering encouragement, this study would not exist. It is to them that I dedicate this book, and to the memory of my father and grandfathers, hired hands in the prairie West.

1

Introduction

In the early 1890s, George Becker pondered his future. He assessed his nature as "being of a type and character who could not entertain to serve others in a servitude position all his life."[1] Like thousands of other like-minded men, he answered the call of the Canadian prairie West: "Arouse up boys, and come to the country where they [sic] can live free and where they will be equal to their masters."[2]

Becker was drawn to the West at a time when prairie farming was still in its infancy. The agricultural frontier of the late nineteenth century was struggling, through trial and error, to establish itself, but it was full of promise. Newcomers were assured that waged agricultural labour would lead to independent land ownership and to a life free from the social constraints and economic uncertainties left behind. After "long deliberation" Becker decided that work as a hired hand offered the best chance to realize his dream to "go on a homestead farming."[3]

From the last decades of the nineteenth century until World War One, men moved to the West by the tens of thousands. In the spring of 1904, Wilfred Rowell joined the throng. He planned to work briefly as a farm labourer, gaining both experience and the cash necessary to start a farm of his own. As an Englishman looking for farm work, he encountered scepticism from prairie farmers who believed that Englishmen were "frightfully lazy and don't even earn their keep." But Rowell arrived on the prairies in the middle of the agricultural boom. In the early decades of the twentieth century, rapid expansion meant that labour was in short supply and high demand. Rowell was able to find work at Penhold, NWT, and soon convinced his boss of his abilities. They agreed upon a year-long engagement. His apprenticeship in farming progressed well. He changed jobs when his boss balked at his wages but found work again at once. He gained enough experience in farm work to begin on his own, and he learned enough about "batching" to be self-sufficient. As a hired hand, Rowell was proud of his accomplishments, secure in his future, and "happy and healthy both in mind and body."[4] The next year he took out a homestead.

By the time Fred Watson came to the prairies in 1913, the boom had ended. He, too, dreamed of farm ownership, but he was unable to reach his goal. His planned brief term as a hired hand stretched into years of short-term jobs on different prairie farms. In 1924 he finally found a longer-term niche at a farm east of Tisdale, Saskatchewan, thankful for the work since it was becoming "more difficult to get a job." He stayed four years.

Watson arrived in the West at a time of agricultural upheaval and consolidation. During the First World War prices soared and agriculture expanded, but in the immediate aftermath prices fell and agriculture contracted. The 1920s were a time of agricultural belt-tightening, with farmers battling the cost-price squeeze as the industry became more firmly enmeshed in capitalist production. For farm workers, the war created extreme labour shortages and, briefly, high wages, but in the aftermath they, too, were forced to tighten their belts. After Watson left his Tisdale job, he was unable to find steady work. Wages dropped and immigrants continued to flood the labour market. In the last year of the decade – 1929 – he could number the days of his work, only "a hundred and two, . . . because of immigration."[5]

George Becker, Wilfred Rowell, and Fred Watson were but three of the thousands of men who lived and laboured as hired hands in the prairie West during the initiation, expansion, and consolidation of the agricultural industry. As part of the parade of newcomers, they both reacted to and helped shape its development. Their experiences illustrate the changing roles and fortunes of this significant segment of the Canadian working class.

The fifty years surrounding the turn of the twentieth century were important in Canadian labour history. The period has been given abundant and careful treatment by labour historians who have documented the experiences of Canadian working men and women as they negotiated their way into the emerging urban-industrial capitalist world. At a time when labouring men and women were both reacting to and effecting important changes elsewhere in the economy and society of Canada, prairie hired hands were also leaving their mark.

During this time, the prairie West was undergoing its own transformation, aggressively altered from an Aboriginal society of hunting, fur trading, and self-sufficient farming to one of a European society geared toward commercial agriculture for an international market. While old, established cities and rural areas in central and eastern Canada transformed themselves within existing institutions and social relations, the prairie West was a frontier both in its material conditions and in the minds of those who moved in. At the same time, the dominant industry of the prairies, agriculture, was undergoing significant technological and ideological change as it passed from pioneer to capitalist modes of production. These conditions and these changes added up to an experience that was unique in Canadian history for a significant component of the Canadian working class. The achievements of workers in the prairie agricultural industry were vastly different from those of their counterparts in central and eastern Canada, yet they held important parallels. Work was central to farm workers' lives, and relationships within the world of work shaped the contours of their social relations.

This study has a twofold aim. It develops a theory to explain the changes in labour-capital relations during the development and maturation of the wheat economy, and it examines the lives of farm workers to understand how they influenced and reacted to these changes. Labour and capital in prairie agriculture appeared to enjoy a relationship of amicability. The lack of overt conflict created the impression of a unity of interest that manifested itself through a co-operative working relationship and a shared ideology in which capital was the only beneficiary. This picture is overdrawn. Agricultural labourers operated within the framework of capitalism but they pursued their own aims, which were often antithetical to those of capital. This was most evident during the half-century in which agriculture became the foundation of the western economy. Between the early settlement period and the onset of the Great Depression, as the agricultural industry developed and consolidated, farm workers had ample opportunity to direct their contribution to the emerging agricultural economy and rural society. They were buffeted by the transformations of capital during this period, but they developed strategies to influence the shape and rate of change in the industry and to maintain significant control over their own working lives. How they did so, and with what varying degrees of success, is the subject of this study.

History is essentially untidy. This study attempts to overcome the difficulties of theme and chronology by employing both. The thematic stage is set in Chapter 2, which establishes the historical, geographical, and theoretical context by examining hired hands and their employers through the nature and particularities of labour-capital relations in prairie agriculture. The study is then divided into three sections covering the broad chronological periods during which the agricultural industry became increasingly enmeshed in capitalist production. In each section, the first chapter begins with a brief description of agricultural developments during that period. The rest of the section then examines the role and experience of farm workers in these developments.

The study begins in the 1870s. The first period, the beginnings of the industry, is covered in Chapter 3. Prior to the turn of the century, the agricultural industry was in its infancy, with new frontiers to be opened and new techniques to be discovered. Farm workers were drawn to the rural West by the great demand for their labour, but more persuasively by a promise that no other industry could offer its workers – independence through farm ownership. The second period, expansion of the industry, is covered in Chapters 4, 5, and 6. Between the turn of the century and the end of the First World War, agriculture developed into a successful commercial enterprise. Farm workers developed strategies to learn agricultural skills, to create a place for themselves in the rural community, and to accumulate enough capital to leave waged labour for farm ownership. The final period, the era of consolidation, is covered in Chapters 7 and 8. During the 1920s agriculture became an established international industry. Farm workers faced their greatest challenges during this period, as their ability to gain farm ownership was eroded and they slipped into a newly institutionalized position as proletarians. Yet even as their position came more nearly to reflect that of workers in other industries, their experiences and responses were shaped by the particular circumstances of the agricultural prairie West.

Within this chronological framework, some themes are repeated and played out as the fortunes of agriculture and the men who laboured in it followed their own direction along history's course. Thematically, the study proceeds along three paths. It explores the nature of hired hands' association with their employers, both as individuals and as actors in the labour-capital relationship in the evolving rural community. It reconstructs the lives of hired hands, examining how their experiences and reactions were refracted through the prism of class. And it situates hired hands in the history of the Canadian working class, both as members of a work force that responded to and effected changes in their industry and as workers arranging their lives around the realities of class.

2

Labour-Capital Relations
in Prairie Agriculture

". . . the rate of wages agreed upon by them is generally decided by the superior bargaining ability, stubbornness or bluff of either party." [1]

Conflict was inherent in labour-capital relations in late nineteenth-century Canada. In 1889, the report of the Royal Commission on the Relations of Labor and Capital in Canada, citing an anonymous observer, reduced the relationship to a single premise: "To treat it (labor and wages) as a simple exchange between equals is absurd. The laborer must sell his labor or starve." [2]

But in the agricultural world of the prairie West the equation was much more complex. An unemployed labourer there did not have to starve. The western prairie was a bountiful mistress, according to popular conception, and any man could at least put a roof over his head and food in his stomach. Indeed, no man had even to be a labourer if he chose not to be. Government generosity and a wealth of unoccupied land meant that any man could leave the ranks of wage labour and become an independent landowner.

Land was abundant, but both capital and labour were in extremely short supply. Acutely aware that an employee of one day might be an employer the next, farmers and hired hands eyed each other as equals. Volatile bargaining strengths led to an uneasy alliance, marked more by accommodation than by conflict. But popular interpretation had more to do with hiding the truism that "the laborer must sell his labor or starve" than did actual prairie West economics. The particular circumstances of agriculture and labour in the West and the two-tier operation of their relationship resulted in contradiction: the mitigation of the basic antagonism between labour and capital simultaneous with an exacerbation of the tension between them.

The Nature of Labour-Capital Relations

Open hostility and stormy confrontation marked relations between labour and capital in the late nineteenth and early twentieth centuries. But in prairie agriculture the relations appeared to be amicable. Hired hands were a valued part of the economy and society, and they were warmly welcomed into farm households and communities. Their economic circumstances were a little straitened, perhaps, but just as secure as any agricultural endeavour could be. Their relationships with their employers were generally good, often excellent. The two shared a common interest not only in the farm's well-being but in the same work routine, the same meals, and the same home. Resistance to poor wages and working and living conditions was usually limited to walking off the job, an action seldom seen as anything more than disgruntlement. While workers in other industries pursued strategies for strengthening their position and improving their lot, agricultural workers appeared content.

The marked dissimilarity from labour-capital relations in other industries reflected the common purpose and interest between labour and capital in agriculture. Labour and capital have fundamentally antithetical aims, but the relationship is complicated, and often mitigated, by time, place, and circumstance, significant considerations in the relations between labour and capital in prairie agriculture. As the basis of the western economy and a principal sector of the Canadian economy, agriculture provided not only the economic impetus but also the ideological rationale for the developing prairie society. The agricultural industry could rely on the practical support of governments and the ideological support of the community, and it retained its political and social importance long after other industries challenged its economic hegemony. This rendered it extremely powerful in its dealings with the labour force and ensured that conflict would be muted.

Even so, the struggle was not one-sided. Labour in agriculture pursued its own aims, and in the process wrested important concessions and influenced the shape and rate of change in the industry. The relationship between labour and capital in agriculture, despite its apparent amicability, was in fact oppositional and conflictual.

It was also complicated. There were important differences between agriculture and other industries and between prairie agriculture and that of other regions. The most obvious was economic. Prairie agriculture involved no ordinary exchange of labour for wages. The line between labour and capital was very unclear; positions often overlapped. Farm workers could be farm owners themselves, or soon to be. They thus played the roles of both labour and capital. Ideology was important, too, in determining the nature of the relations. In prairie agriculture, farmers and hired hands both perceived the opportunity to move from labour to capital as a realizable goal. Farm workers did not simply take a job; rather, they entered a community with intricate and subtle socio-economic dimensions.

Prairie farm workers found themselves in a complex situation, seen most clearly in the apparent dichotomy between an abstract or theoretical perception and a concrete or practical application of the relations between labour and capital. In prairie agriculture, the line separating perception from reality is never very clear-cut, but the relations can be understood if they are studied at two levels. The first is theoretical, that is, the general relationship between labour and capital in agriculture; the second is empirical, how these implications were worked out for the men involved. Here we have a further subdivision. Labour and capital in the agricultural industry are represented in the aggregate by the work force and the owners of the means of production, and individually by hired hands and farmers. It is at this individual level that we can see and define the particular relations that developed between hired hands and farmers on prairie farms.

The distinction between the theoretical and the empirical is important. While the industry itself operated according to capitalist economic imperatives, the men involved, whether employers or employees, were motivated by non-economic considerations as well. They wanted to farm. They aspired not to become capitalists who derived their wealth from the mere ownership of the means of production, but to be working operators of the farms they owned. The distinction between the theoretical and the empirical is thus not a dichotomy. Farmers who worked as farm labourers as well as hired hands who hoped to become farmers sought their success within a capitalist framework. The result was an uneasy peace. Farm workers and their employers did not confront each other as pure labour and pure capital. Rather, their common aspiration to an intermediary position between labour and capital muted the differences between them, and they saw their interests as converging rather than diverging. This is one key to the relations between labour and capital in prairie agriculture, but it does not alone explain the particularities of the relationship.

The foundation of the relationship was laid just as the industry itself was being formed and a new prairie society was being established. This gave inordinate influence to conditions at the inception of the development, both to the shape of the relations themselves and to the unevenness of subsequent adjustments to changing circumstances.

Historical Background

To understand how this situation came about, it is necessary to explore the complex interplay of central Canadian purposes in developing the prairies and the federal policy for the western lands. The economy of the British North American colonies in the 1860s was insecure. Britain had withdrawn her mercantilist protection, and the 1854 reciprocal trade agreement with the United States had collapsed. To an important group of central Canadian capitalists the West lay waiting, ready to provide the markets necessary to ensure the continued prosperity of central Canada.[3]

The acquisition of a territory five times the size of the rest of Canada presented the opportunity for a massive agricultural industry. Yet despite glowing federal propaganda about the fertility of western Canadian soil, a large portion of the region was still condemned as semi-arid. Palliser's Triangle, a block of 16,000 square miles of prairie centred on the present-day boundary between southern Alberta and Saskatchewan, had been assessed in 1859 by Captain John Palliser as unfit for cultivation, the northern extension of the "Great American Desert." Henry Youle Hind, commissioned by the Province of Canada, agreed with this assessment but painted a glowing picture of the wide belt of parkland that bordered it on the north. It was to this area, extending from the Red River northwest along the North Saskatchewan River, then heading south in the shadow of the Rocky Mountains to form a wide fertile arc across the prairies, that expansionists in central Canada turned their attention. The size and fertility of this area grew steadily in the minds and in the literature of the promoters of western settlement.[4]

Throughout the 1860s, the soil was reassessed as among the richest in the world and the climate and topography as ideal for agriculture.[5] When botanist John Macoun carried out exploratory surveys during the 1870s, his reports of enormous agricultural potential were welcomed.[6] Hind vilified Macoun and his findings as the "one-sided speculations of an incompetent amateur or grosser perversions of an unscrupulous charlatan,"[7] but Macoun's views prevailed. The new optimistic impression was as unrealistic as the old pessimistic one had been, but by the 1870s even the northern part of the prairies had come to be regarded as "the future garden of the west."[8]

This perception helped lay the foundation for the West's particular economic development. Increased primary production would help fuel the economy, and a larger economic base would result in a stronger economy. The emphasis was on scale of production rather than on diversification or economic independence. Just as Canada had been part of the colonial empires of France and Britain, the West would be the colonial empire of central Canada, producing raw materials for export and providing markets for goods and services from the centre.

Westward expansion provided tangible political benefits as well. Capitalist support for a government providing such a lucrative market was the most obvious political reward. Industrial workers, promised more jobs and security through expanded markets, were enthusiastic in their support. Ontario farmers regarded the West as a patrimony for their sons unable to take up farms in an overcrowded province. Beyond a simple collection of votes, the acquisition of the prairies would ensure Canadian sovereignty in the northern part of the continent, protecting it from America's grasp.

Out of this economic and political interplay came a comprehensive plan for western development. An agrarian economy and society were to replace the fur trade. Small-scale, family-sized units of production would provide a market large enough to absorb central Canadian manufactured goods. Single-crop production would provide the volume of grain necessary to establish secure international markets. Small landowners and their families would provide a conservative foundation for a stable population. The agricultural industry designed for the West was to provide the basis not only of the economy but of the society as well.

A plan was devised to carry this out as rapidly as possible. In 1872, the Dominion Lands Act followed the lead of the United States in encouraging agricultural settlement and expansion.[9] Canada offered 160 acres of prairie land to any adult male willing to wager ten dollars that he could bring forty acres under cultivation and erect a building or two within three years. Because the policy had a dual purpose – to establish both an agricultural economy and an agrarian social structure – compromises were necessary to ensure that both aims were fulfilled. As a result, the homesteading policy was geared more to rapid and massive population expansion than to efficient agricultural production. The low filing fee attracted settlers who lacked the capital to buy the equipment necessary for grain growing on a large and efficient scale. Homesteading obligations of cultivation, construction, and residence required to "prove up," or gain clear title to the land, were continuously adjusted to allow cash-poor farmers to spend long periods of time away from their own farms.

The effectiveness of the Dominion Lands Act as an instrument of western settlement has been considerably studied.[10] But the Act's influence on the relations between capital and labour in prairie agriculture has yet to be considered. The system of 160-acre free homesteads and the regulations both created and solved some of the problems of labour demand and supply. Critics of the scheme correctly argued that the giveaway of western lands would drive up the price of labour.

From the outset, farmers constantly complained that they were unable to afford or even to find hired help. But the promise of free land was critical to attract labour. Without the hope of eventual ownership, there would have been no labour force at all.

The synthesis of labour and capital thus resulted in a shortage of both. The rationale behind establishing the agricultural industry in this manner was the assumption of the traditional model of small-scale agriculture in which owners are their own labour force. Agriculture under these conditions has limited potential for growth, but the West was not prepared to accept any limits. The contradiction in the agricultural industry was that despite its small-scale organization, it was filled with ambitions for large-scale production. In most industries, such grandiose plans can be achieved through greater appropriation of the surplus value of labour. But in the agricultural West, labour would not comply.

There was little economic incentive and limited opportunity for men to work full-time as agricultural labourers. Even large-scale commercial farms could not compete with other primary industries for labour, and the emphasis on grain crops meant that most employment in agriculture was restricted to the seasons of planting and harvesting. On the other hand, most men neither needed nor sought full-time work since they could file for homesteads. The small filing fee meant that men with very little capital could take up land. The costs of the first year – seed, implements, provisions, putting up a shack to live in – could be earned by working for wages for an established farmer. Homesteading regulations made explicit provision for this; a homesteader was never required to live on his quarter-section for more than six months of each year. Immigration pamphlets spoke in glowing terms of employment opportunities available to undercapitalized homesteaders.

The ideological contradictions this system generated were striking. Free land gave labour the opportunity to be extremely independent, but at the same time it created a landowning class of small-scale capitalists. Often, the two roles coincided. A farmer might alternately be employer and employee, working on another farm during the summer once his own crops were sown and hiring extra help at harvest. He occupied the positions of both labour and capital and faced the contradiction of playing dual and oppositional roles. It was a circular process: the availability of land that caused the labour shortage in the first place undercut the independence that could have resulted from the shortage of labour. In principle, during the formative period, labour was invisible, or at least it did not exist as a separate category because the industry owners were also its labour force.

This type of labour force provided distinct advantages to governments as well as to railways, manufacturers, and suppliers. As owner-operators of small units of production, the homesteaders could be relied on for political stability. Unlike transient agricultural workers, whose characteristic seasonal unemployment created hardships for themselves and headaches for governments, these workers went quietly back to their homesteads once they finished their work. Their vested interests in the region could translate into rapid and orderly development of social institutions. Their needs for goods and services were much greater than those of mere wage-earners. They were engaged in much more than simple survival; they were building farms and communities, as well as a niche for themselves in the prairie West. Labour in agriculture was thus expected to play a significant part in the social as well as the economic development of the agrarian West, but not in its role as labour. Unlike other industries, in which labour must be coerced into supporting the interests of capital, labour in agriculture was expected to *be* capital, either in the present or in the foreseeable future.

Special Conditions of Prairie Agriculture

Labour-capital relations were also shaped by three particular characteristics of the prairie agricultural industry. The method of production, the nature of the product, and the organization of the industry all created conditions that were quite unlike those in other industries.

Agriculture is a mass-production industry in that raw materials are transformed into finished products on a large scale, yet it does not enjoy the controls and precision of factory production. It is a resource extraction industry in that natural products are harvested, yet these products do not occur naturally but must be carefully tended and continuously replenished. To be commercially successful, the industry must employ large-scale, efficient methods of production while adapting itself to the many environmental variables of production beyond its control.

Agricultural products are unlike those in most other industries, for they do not grow according to laws of supply and demand. It is both the charm and the bane of farm products that they need long sunny days, sufficient rain at just the right moments, and favourable weather as the crop ripens. Seasonal variations are an integral part of agriculture, and capital's response to seasonality is to make extremely fluctuating demands on labour. In a grain-growing economy labour demands are low for most of the year, but for short, very specific periods the demands are urgent and high. Agricultural diversification can moderate the seasonal rise and fall in labour demand. Mixed farming, with a variety of crops, spreads the planting and harvest seasons over a broader period, and the production of livestock requires daily care, levelling off the peaks and valleys in labour demand.

Nonetheless, seasonality remains one of the most striking features of all agriculture.

As winter ends and the soil thaws, the land must be ploughed and sown to a crop. The short growing season in the West dictates that this be done quickly to take full advantage of every growing day. Spring planting offers workers their first regular employment of the year. The length of this period depends on the weather and on the variety of crops; in prairie grain-growing, it rarely extends beyond a month. Through the summer, labour requirements remain low but fairly steady. There is always farm work to be done: cultivating the growing crops, repairing buildings and equipment, caring for livestock, cutting brush or breaking new ground, ploughing fallow fields. Late summer sees another rush as the crops begin to ripen. Harvest brings the highest and most urgent demand for workers. The early frosts of a prairie autumn demand that crops be harvested during a very brief period. Fall operations are divided into two parts: the actual harvesting or cutting of the crops, then the threshing, removing the grain kernels from the stalks. Good weather is vital; a wet spell or an early frost is a disaster. For the labour force, harvest work is easy to find and harvest wages are high, higher than those in other industries and sometimes high enough to bankroll workers for the coming winter. After the crops are cut, there is less urgency. Once bound into sheaves, the grain can be left to ripen and be threshed under a less demanding schedule. Grain unthreshed before winter can be stacked under cover to protect the sheaves from rain and snow. Labour requirements and wages are somewhat lower, but this is an important period for workers, the last chance to fill out the winter's stake. Once the crops are in and after freeze-up, there is seldom any need for hired labour. For those who spend the winter doing chores, wages commonly consist only of room and board and a bit of tobacco money. With a new spring, the cycle begins again.

It is in the method of agricultural production that the overlap in the relations between labour and capital is particularly striking. For the labour force, what is important is the work process itself: the actual tasks performed and the level of skill needed to perform them. The labour process adds both a dimension in which labour can exercise some control and another area of conflict in labour-capital relations.

Work in agriculture is extremely varied, usually physically demanding, and repetitious, but often challenging and fulfilling. Agricultural tasks range from the backbreaking boredom of picking rocks to the finesse of caring for brood sows, from long dusty days ploughing fields to fragmented days filled with numberless chores, from the isolation of stringing a fence to the teamwork of harvesting a crop. The skills required range from nothing more than physical strength and stamina, as when clearing brush or shovelling out the stables, to highly developed expertise, as when handling a recalcitrant team of horses or repairing delicate machinery. Yet agricultural work is not highly specialized. The labour force is expected to perform each and every task. At the same time, the necessary skills are not the exclusive property of farm workers but also the province of farm owners. Unlike other industries in which labour is hired to perform tasks the owners or managers of capital cannot perform, in agriculture the employers are able to do the work themselves.

Labour and capital in prairie agriculture thus shared many of the same skills and performed the same tasks. There were exceptions, of course, and examples of farm hands instructing their employers, but by and large farm hands were expected to reproduce the skills possessed by their employers. Capital sought a labour force that could duplicate its own skills rather than provide new or specialized skills, and labour was hired not to provide basic production but to increase productivity.

The organization of the industry further moulded the relations of labour and capital. On the prairies, as throughout Canada, agriculture was based on the small owner-operated unit of production. There were large farms and there were farms that owners did not operate, but they were the minority. During most of the year, the owner-farmer and his family could meet their labour requirements. Moreover, particularly in western Canada, farmers were capital poor, which meant that they frequently worked at waged labour themselves and that, when they did hire a worker, they usually did not hire year-round, full-time help. The few who did seldom engaged more than one hired hand. The broad result was that the ratio of labour to capital was very low, and the proportion of the agricultural population represented by waged labour was also very low. Employers in agriculture greatly outnumbered their workers.

The situation for labour in prairie agriculture was thus quite different from that of other industries. Although capital most frequently holds the upper hand in its dealings with labour, it must nonetheless make concessions because it relies on labour's skills and numbers. In prairie agriculture, labour's skills were not its exclusive property, and workers could not often rely on their numbers for strength.

The authority the industry wielded made labour's task even more difficult. As the dominant prairie activity, farming extended beyond the economic sphere to influence the social and ideological spheres as well. For agricultural settlers the West was a new land, to be fashioned to specifications they harboured in their residual cultures and adapted to conditions they encountered. The economy and society were evolving not only simultaneously but conjointly, giving particular strength to the new institutions and to the agricultural industry upon which they were based. To complicate the situation for labour, to a much greater extent than in other industries, economic and social dimensions in the relations between labour and capital merged. There was seldom a clear distinction between the two, and their overlap helps to explain why farm workers acted as they did.

Prairie rural society was based on the family farm as an economic and social unit, and the family farm created much stronger social and economic links than those in other industries. The farmer lived on his farm, and so did his hired help. The agricultural work force thus shared more with its employers than skills – it usually shared the same food, the same roof, and when necessity dictated even the same bed. Moreover, hired hands were required to accommodate themselves not only to their employers but to their employers' families. The majority of their work contacts were made with their employers and their employers' families rather than with fellow workers, and so, too, were most of their social contacts. Because few farmers engaged more than one hired hand, farm workers were isolated not only from one another but from other working-class groups as well.

Another dimension in which the agricultural industry exercised a less direct but no less powerful influence was the milieu of the agrarian society. The social importance of agriculture matched its economic weight. In sheer numbers, the proportion of the population engaged in agriculture, both directly and indirectly, was the majority. In the larger context of social relations throughout the West, the farmer was no match for a mining magnate or a large-scale rancher. But in the rural community he occupied, he was paramount in his position as a landowner. This was an agrarian world. Farmers were far from a cohesive group themselves, but they and their families were arbiters of society, determining the mores and the culture of the rural prairie West. Churches, schools, and other social institutions fostered agrarian social values. Not only farmers and their families but other segments of the rural population that relied on agriculture for their livelihood and on the agrarian community for their social sustenance embraced these values. Farming was regarded as more than an economic endeavour. It was a way of life that provided the underpinnings of an entire social system.

Farm workers did more than simply work. They entered a social milieu with a particular ideology that held a very ambiguous place for them. Labour's small size, both relatively and absolutely, and the blurring of the lines between labour and capital resulted in strong pressures. Hired hands were expected as a matter of course to act in the interests of capital rather than of labour.

The Agricultural Work Force

In prairie agriculture, capital held ultimate economic and social authority, yet labour did much to determine the nature of the relationship. Farm workers were situated within the broad context of labour in the West and at the same time were influenced by the specific conditions of developing prairie agriculture. They belonged to a work force characterized by occupational plurality and locational mobility. However, the men who chose to enter the agricultural sector of the work force fit more closely, but not exclusively, into a more permanent category.

Between 1880 and 1930, as the agricultural economy of the West was established and developed, farm workers made up the largest segment of the waged work force on the prairies. Of these, the majority were full-time hired hands. It is difficult, particularly in the early years, to separate the men who worked primarily for farm wages from those who derived most of their income from their land. The difficulty is compounded because the early labour force was fluid, as men shifted easily between waged agricultural work and work on their own farms.

The agricultural labour force on the prairies was made up of three components. The largest single group was farmers, whether owners or tenants of the farms they occupied. The second and third groups consisted of unpaid family members and paid agricultural labourers. The relative proportions of the three groups shifted over the five decades of this study, with an overall decline in the percentage of farmers and an increase in that of waged workers. In 1891, waged farm workers made up 14 per cent of the agricultural work force. Their proportion grew to 16 per cent in 1911 and to nearly 19 per cent by 1931.[11] The increase was slow and steady, reflecting a work force that contained a relatively stable but growing proportion of hired farm workers. But among the individuals, there was a regular turnover.

The hired hands of this study were wage workers regularly employed as full-time permanent agricultural labourers on mixed and grain farms.[12] "Permanent" does not mean that agricultural work was the only employment a man might undertake in the course of a year. But if a "permanent" hired man sought work outside agriculture, it was only a temporary expedient. Nor does "permanent" suggest that the farm worker occupied an unchanging niche. Farm work for wages was seldom intended to last for a man's working life. "Permanent" distinguishes regular hired hands from temporary farm workers such as homesteaders who worked part-time for other farmers, or seasonal workers such as harvesters who joined the agricultural work force only for short periods.

The term "full-time" is likewise problematic. It refers to the nature of the work performed, more specifically to the type of jobs an individual farm hand might obtain. It does not indicate that farm jobs were long-term. Men designated as "full-time" workers were those employed as eight-month or year-round hands. Their jobs might extend over several years or even longer. Equally, they might undertake a number of short-term jobs on several different farms or have more-or-less regular positions with one farmer for specific periods and for jobs such as spring planting or winter chores, and spend the rest of the year working on other farms throughout the district. The term is used here to indicate that these men regarded farm work as their main, sometimes their only, type of employment and their major source of income.

But not all men who were hired hands fit this description so neatly,[13] and many men hovered on its periphery. Some specialized types of agriculture, such as sugar-beet growing or dairying, were labour-intensive, and the men who worked in these industries experienced a different workplace and developed different relationships with their employers.[14] Similarly, harvesting drew large numbers of men who were prairie agricultural workers only for this seasonal occasion. During the harvest season, when temporary labour requirements far outstripped the available pool of labour on the prairies, men flocked from other western industries and were imported by the thousands in annual government- and railway-organized harvest excursions from central and eastern Canada, British Columbia, the United States, and even Great Britain.[15] Although they shared a similar work environment with hired hands, the nature of their involvement in the agricultural industry was vastly different, and so, too, was their relationship with their employers and with permanent farm workers.

Despite the complexity of the agricultural work force, men who undertook farm work had distinct ideas about their location in it. The way they answered the census enumerator makes this clear. It is reasonable to assume that most men who identified themselves as agricultural labourers were permanent full-time farm workers. In an economy and society based on independent farm ownership, men who could claim to be something other than farm workers would have been proud to do so. Farm owners or tenants who were working on another farm during the enumeration would not have identified themselves as labourers. Given the higher status ascribed to farming, they would have described themselves as farmers, even if they were tenants or homesteaders waiting to prove up. Despite all these qualifications, however, for the purposes of this study the men who were defined as paid agricultural labourers in the census will be those who are described herein as "hired hands." Such a definition encompasses their varied circumstances and anchors them in the world in which they worked and lived.

BEGINNINGS, 1870s-1900

3

Recruiting the Agricultural Labour Force

"[The prairie West is] as good a place as a man can find if he has plenty of money and brains, or if he has no money but muscle and pluck." [1]

In December, 1874, Albert Settee signed a contract "hereby agree[ing] to work for Colin Inkster as a farm laborer for the space of Four Calendar Months" during the following summer season. For "the faithful performance of such service" he was to receive the sum of four pounds per month.[2] Such written contracts were rare. This one demonstrates a number of features in the relationship between labour and capital during the initiation of the agricultural industry.

Settee signed the contract in mid-winter, a time of the year when farm work was scarce, if not impossible to find. The assurance of employment for the following summer may well have been all that enabled him to survive the rest of the winter. His willingness to agree to work at the same rate of wages for the entire summer, including the traditionally higher-paid harvest season, reveals his need for a secure job. Yet the advantage was not all to his employer. If Settee needed a steady job, Inkster just as surely needed a steady man. The contract reflected an anticipated uncertainty of labour supply. It ensured that Settee would stay in the district the following year and that Inkster would not have to scramble to find hired help. The advantage to Settee would be more evident the following year when he collected his wages. Four pounds per month was not high in comparison with wages for other labouring jobs in the West, but room and board were included and the work was certain.

Both men were gambling on the weather. The contract ran from the beginning of May, and if Inkster was planning to wait for Settee to begin spring seeding, he was taking a chance that the crop might not ripen before the usual mid-August cooling that slowed grain maturation. If summer weather was cool, the harvest might extend into September, giving Settee a distinct advantage whether he bargained to stay on with Inkster at the going harvest rate or chose to seek work elsewhere in the district. A warm season, though, could result in a harvest as early as mid-August and would leave Settee tied up until other high-paying harvest work in the district was completed.

The mutual benefits and drawbacks that Settee and Inkster found in their contract reveal the challenges an industry faced in becoming established. The scant population meant that labour was in short supply. But the nature of agriculture, particularly cereal monoculture, meant that labour was in high demand for only part of the year. The low wages indicated that capital, too, was in short supply, but the ready availability of land as a substitute offset this. The major problem confronting the agricultural industry was a shortage of both cash capital and labour. By using land as capital, the industry was able to secure a labour force instilled with expectations that dovetailed neatly with the industry's own needs. As for the labour force, the combination of seasonality, abundant land, and the ability of the worker to become an owner created a much more complex relationship than existed in any other industry.

The short term of the contract illustrates not only the wage-determined and seasonal nature of agricultural labour needs, but the short-term nature of labour-capital distinctions. By the end of 1875, Settee may well have been working on his own farm, and Inkster may once again have been seeking an employee. Although it was uncommon for labour and capital in prairie agriculture to formalize terms of employment, as did Settee and Inkster, the circumstances that caused them to do so reflect the broad conditions of prairie agriculture during the birth of the wheat economy.

Pre-1900: Agricultural Beginnings

When Albert Settee and Colin Inkster signed their farm labour contract in 1874, prairie agriculture could scarcely be said to exist as an industry. It was not until 1876 that the first commercial shipment of wheat left Manitoba: 857 bushels travelled by riverboat, rail, and steamship to Toronto. The entire amount had been grown by twelve local farmers in lots ranging from 17¾ to 204 bushels and fetched them 80 cents a bushel. The shipment was small, but the *Manitoba Daily Free Press* announced grandly that it was "fraught with the most important results to the agricultural and business interests of the North-West."[3] It signalled the beginning of what would become one of Canada's most important industries. Two years later, a consignment of grain followed the same route to St. Paul and then by rail to the seaboard, to be transferred to a freighter for delivery in Glasgow.[4] Prairie agriculture had taken its first steps toward becoming a huge, profit-making enterprise, incorporating technological innovations, producing for international markets, and consuming large amounts of both capital and labour.

Before the 1870s, agriculture had been peripheral to the western Canadian economy. Production was self-sufficient or had been geared to local markets, and was confined to the Red River colony and the areas immediately surrounding the trading posts.[5] Yet by the end of the century, agriculture was the mainstay of the western economy. From the diversity of mixed farming for local consumption to the specialization in a single cash crop for export, agricultural energies became directed to the production of wheat.

But the rewards for such single-mindedness were still in the future. Throughout the closing decades of the nineteenth century, the fortunes of commercial wheat production varied between stagnation and spurts of growth. The period was characterized by frustrated optimism, anticipation rather than realization of the agricultural potential of the prairie West. The realities of remote and demanding markets, an insufficient and costly work force, and a harsh and unpredictable climate tempered high expectations for rapid expansion.

Farmers moving onto the Plains soon found they needed new agricultural techniques. It was the uncertainty of rainfall – the problems both of insufficiency and of bad timing – that most plagued the industry. Fallowing was finally seized upon as the best method of storing for next year's crop every drop of moisture the dry prairie climate doled out. By the 1890s the value of fallowing was also evident in weed prevention and soil renewal. New techniques of cultivation, tillage, and seeding met other climatic challenges, such as dry prairie winds.[6] The difficulty, though, was that manpower requirements were heavy. Operations such as ploughing under the stubble rather than burning it or cultivating and harvesting a seed plot by hand were labour-consuming refinements few farmers could afford. Instead, they concentrated on volume.

As in most other industries in the late nineteenth century, agriculture increased production through mechanical and technological improvements. By the time the West was opening to agriculture, even though much of the old equipment was still in use, farm implements and equipment were reflecting rapid technological improvements.[7] In the 1820s, with traditional hand agricultural tools and horses, three man-hours of labour were required to produce one bushel of wheat. By the 1890s, largely due to mechanization, the same bushel of wheat could be produced in one-half man-hour.[8] Without such mechanization, commercial agriculture on the prairies would have been impossible.

Initially, the primary aim of mechanization and improvements to agricultural equipment was not so much directly to reduce labour requirements as to make more efficient use of labour's time. Improvements to the plough shortened the time to prepare the soil for seeding. One man working with a team of oxen or horses and a steel walking plough would be hard-pressed to plough even three or four acres per day; at that rate, it would take years before a quarter-section grain farm was producing to capacity. A sulky plough provided a seat for the driver, but did little to speed the work. More important was the gang plough, an implement that strung together a number of ploughs in order to cut more than one furrow at a time, bringing more acreage under cultivation quickly.

Harvesting the larger crop also called for more sophisticated machinery. Using a cradle, a man could cut and bind no more than one acre a day. The horse-drawn reaper, which could cut six or seven acres a day and leave it in piles for binding by hand, provided the much greater speed needed to harvest large prairie crops. But hand binding was slow – five man-hours were needed to bind an acre of wheat.[9] In 1877 an American farm worker invented a device to tie these piles into sheaves. Behind a good team of horses, these new self-binding reapers were capable of cutting and binding fifteen to twenty acres a day.

New technology supplemented mechanization. Oxen and horses were the major adjunct to human labour, but as early as 1874 the steam engine appeared on the Canadian prairies.[10] Steam could not entirely replace horsepower for tillage, but it did allow crop production to increase dramatically. Steam engines were capable of pulling ploughs that could turn twelve or fourteen furrows at once, enabling much larger acreages to be brought under cultivation rapidly. But it was in threshing that the steam engine was most important, replacing horses as the power to drive the threshing machines. Threshing was not subject to the same constraints of time as harvesting because prairie farmers cut grain before it was ripe, then allowed it to ripen in the sheaf. But as wheat harvests grew, steam engines had to thresh seventy-five acres of wheat in a day to keep up with production.

Rather than reducing manpower needs, the new machinery and technology increased them by allowing greatly expanded production.

The horse-drawn reaper required only two men to operate it – one to drive the reaper and another to throw the cut grain into piles for later binding. The self-binding reaper could harvest as much as seventy-five or eighty acres a day but needed a crew of three or four men to follow after it, piling the bound sheaves into carefully constructed cone-shaped stooks. Steam threshers needed crews of eighteen to twenty-five men. By the 1890s, agricultural production outstripped the labour supply, and western farmers were calling on work crews from central and eastern Canada to help bring in their crops.

At the same time, though, anticipated success eluded the industry. Fluctuations in the actual volume of grain production forcibly indicated the uncertainties plaguing the industry, despite a steady increase in the number of farms and in acreage under cultivation. Between 1881 and 1901 the number of prairie farms increased more than fivefold, and improved acreage increased more than twentyfold.[11] From 54,000 acres planted to wheat in 1880, wheat acreage reached more than one million acres eleven years later. By the end of the century it reached almost two million.[12] But production figures did not mirror steady growth.

Although the proportional increase in prairie wheat production between 1880 and the end of the century was spectacular, from just over one million bushels in 1880 to 63 million bushels in 1901, detailed examination paints a less rosy picture. A Manitoba crop of less than 6 million bushels in 1886 doubled the next year to more than 12 million, but by 1889, Manitoba production had fallen back to close to 7 million bushels. A year later it doubled again, to 14.5 million bushels in 1890. The following decade saw even wider shifts. A bumper Manitoba crop of more than 23 million bushels in 1891 was followed by a three-year series that averaged less than 16 million bushels annually. But the 1895 crop, at more than 31 million bushels, was the largest of the century. The very next year saw the figures fall back to 14 million, then rise again to almost 28 million in the last year of the decade. The future did not hold much promise of improvement. The first crop of the new century was the lowest since 1892 – only 13 million bushels.[13]

Population growth in the closing decades of the nineteenth century was likewise irregular. It was also disappointingly slow. The Immigration Branch of the Department of Agriculture launched an immigration drive in Great Britain, Europe, and the United States, and the Dominion Lands Act of 1872 was expected to draw a flood of settlers. When only a trickle appeared, hopes were pinned to the completion of the Canadian Pacific Railway in 1885. That wish proved elusive as well. The uneven fortunes of agriculture were only one of several reasons for the slow population growth in western Canada but it was significant, holding back potential settlers and precipitating the outmigration of those who failed.

Despite the vagaries in wheat production and the slow rate of settlement, the agricultural industry did achieve overall growth. The key was expansion. Increase in wheat acreage was the most important single indicator of the development of the wheat economy, demonstrating that continued growth could come about through continued expansion. In the closing decades of the nineteenth century, it appeared that the prairie West had all the requisites for a successful agricultural industry save one – labour.

Casting the Net

"I arrived in Winnipeg, July 1881, with only one sovereign in my possession," reported an immigrant from Dublin in 1883. Before two years had passed, he could boast that "I have a quarter of a section of land (160 acres) within half-a-mile of the railroad, and I have a house built upon it and some ploughing done." He believed his prospects were bright. "I will be able to hire a man next year," he declared confidently, "when I hope to have good crops, and a new house built on my land."[14]

The route to this achievement was employment as a hired hand. The Dubliner faced no difficulty finding farm work. Upon his arrival in Winnipeg, he "immediately went to Mr. Hespler, the Canadian Government Agent, whom I found a very nice man indeed." Hespler sent him to a farm at once, fifty miles from Winnipeg, where he demonstrated his mettle and "buckled to work." Within "a short time," he moved on to the North-West Territories and selected a homestead. But he found that a shortage of cash slowed his progress: "If I had some capital I should have had a fine crop this year." He realized that his days of farm labour were not as brief as he had expected, but this discovery did not dampen his enthusiasm. "I have been obliged to work out for some time," he explained, "but next year I hope to have a good crop." More importantly, he would join the ranks of men who had moved up from waged farm employment to the independence of farm ownership and the rank of farm employer.

Ambitious, independent men saw themselves in the Dubliner's tale. His assurance that his own rapid advancement was based solely on hard work struck a responsive chord among men who faced shrinking or closed opportunities at home. If they, too, were willing to work diligently, should they not be equally rewarded? If men like the Dubliner, of humble origins and few material resources, could become prosperous farm owners, there was nothing to hold back the thousands like him who were not content merely to maintain their economic position when hard work and resolution could bring them independence and success as the full fruit of their labour.

Their ambitions coincided with imperatives for economic development in the West. Canadian government and immigration agents, in conjunction with colonization and transportation companies, stepped up their efforts to capture the migrating European and North American populations and funnel them to the West. They cast their advertising net at agricultural settlers.[15] The lure was free land. For promoters of an agricultural West, the vast prairie lands could be substituted for capital as the reward for the labour necessary to establish the industry. For newcomers, economic and social improvement seemed a fair exchange for hard work and fortitude. The early pioneering period was thus a time of promise.

Central to the fulfilment of the promise was the assurance that farm ownership would provide independence. The most tangible sign was material improvement, the degree of which was said to depend on an individual's capacity for hard work. Less tangible measurements stemmed directly from the promised financial self-sufficiency and included an improvement in social status that could extend even beyond social equality to social leadership within the community. Ultimately, personal contentment, social standing, and economic security beckoned.

Immigration propaganda was thus explicitly directed at men whose ambitions stretched beyond simple farm labour but whose economic circumstances might compel them to undertake it temporarily. It was prescriptive in soliciting men who had "no money but muscle and pluck"[16] and actively sought those who would make a vital contribution to agricultural growth, both through their labour and through their investment into the land of what capital they did acquire. It was proscriptive in warning off shirkers. The West "is not the place for people who cannot make a living in their own home," declared farmer W.M. Champion, "but any man of strength and energy can do well here."[17] The literature of the period is sprinkled liberally with calls for men with "strong hearts and willing hands" who were "steady," "sober and industrious," men with "energy" and "perseverance" who could "make light of discomforts."[18] It sought men who aspired to be working farmers, who were not expecting an easy ride to wealth. Intending immigrants were warned that it was a "great mistake" to think "they need only to stoop down and pick up the lumps of gold lying by the roadside," but they were assured that "whosoever is able and willing to work will be well repaid for it."[19] These criteria had been determined even before the West was opened, as Doug Owram has shown in his study of the Canadian expansionist movement: "the ideal society envisaged for the future was one based on the independent rural landowner. While he was unlikely ever to become rich, it was thought he could become comfortable by working the land."[20]

Contemporary observers believed that intending settlers would want reassurance that the North-West could provide "a resting place of comfort, of independence and freedom."[21]

Newcomers who wanted more from their land frustrated the plan for a yeoman society. The "development and sale of land for a profit was relatively common," according to R.G. Marchildon and Ian MacPherson. "Many speculated in ways that would have warmed the heart of any aggressive Winnipeg real estate salesman."[22] But such speculators were unwelcome both to local farmers and to governments, as homestead regulations indicate. More often the men who took their cue from the promises were those who were willing to work hard and to live frugally, and who fully expected to win the reward of a comfortable independence.

Expert voices enhanced the official endorsements of governments and the blandishments of private land and transportation companies. The Department of the Interior and the Canadian Pacific Railway employed in the propaganda campaign authorities such as John Macoun, who had been rewarded for his services in reassessing the agricultural potential of the prairies with an appointment as Dominion botanist, and Alexander Begg, a journalist, rancher, and ardent promoter of settlement in western Canada. At a time when farming was so difficult and the prospects of successful agriculture so uncertain, their testimonies were expected to act as correctives to the unfavourable publicity filtering back to prospective immigrants, and to entice the doubtful to the prairies. Macoun was already well known for his extravagant assessments. He declared that the North-West contained the "Best Soil in the World," hinting that the temperatures of New Orleans were visited upon the southern prairies "and confer on it the blessings of a climate, not only exceptional as regards character, but productive of results to the agriculturalist, which . . . are unsurpassed in any other part of the world."[23] Begg waxed enthusiastic about the rewards awaiting settlers. "Manitoba and the Canadian North-West are . . . essentially agricultural countries," he insisted, "and it is to the tiller of the soil that they offer the greatest inducement. Farmers from the old country, and those who have a knowledge of farming may, with care and industry, prepare for themselves a future of independence and comparative ease."[24]

Even more effective were first-hand accounts of established settlers. Tales of achievement such as that of the Dubliner received widespread publicity. Macoun backed up his claims with the advice of successful farmers. "I would advise any young man with good heart and $300 to come to this country," he quoted farmer Joshua Appleyard of Stonewall, Manitoba, "for in five years he can be independent."[25] Begg's use of settlers' declarations was even more refined. In 1881, the Canadian Pacific Railway appointed him as General Emigration Agent to publicize the potential of the western prairies. Through the distribution of questionnaires in 1884 and 1885, he solicited the testimonies of pioneer men and women about their settlement experiences and published a number of pamphlets culled from the positive replies.[26] The Department of Agriculture endorsed the project by joining the Canadian Pacific Railway in sponsoring the publications.[27] Begg found many examples of satisfaction and economic improvement. Most of the respondents reported that they were at least comfortably off, with farms valued from a few hundred pounds upward.[28]

The most successful farmers started with some capital, and there was universal agreement that money would provide an easier start. When British observer Henry Barneby published *Life and Labour in the Far, Far West*, he established his credentials by noting that he was "indebted to Canadian residents" for his information. "People have told me," he informed his readers, "that they consider a settler should have at least enough money to keep him in food for two years." He suggested that 330 to 400 pounds would provide a good start for two people:[29]

Yoke of oxen, say at Qu'Appelle	£50
Waggon	6
Plough	5
Farm tools, say	20
One year's supply of food for self and wife (and this is a low estimate)	60
Lumber for house and stable, for building a four-roomed house	60
Two cows, say	30
Journey out for two, say	40
Extra cash for seed, &c., and contingencies.	
Homestead fee, 160 acres	2
160 acres pre-emption land, at 2 1/2 dols. per acre	80
	£363

Barneby's list of expenses was realistic, but countless farmers testified that they began with much less. The estimated amount needed to begin farming varied, depending on the degree of hardship a settler was willing to endure. Five hundred dollars was a figure often cited in immigration literature, but the Canadian Pacific Railway assured single men that $385 would give them a good start.[30] Less than $300 would condemn the homesteader to "a life of penury," according to the Reverend Nestor Dmytriw, an observer of Ukrainian settlements in Canada in 1897.[31] Lyle Dick confirms these figures in a study of farm-making costs in southern Saskatchewan from the 1880s. In a challenge to a study by Robert Ankli and Robert Litt, which estimates that settlers at the turn of the century needed a minimum of $1,000, Dick has concluded that a settler could begin farming with only $300 to $550, "albeit in a rudimentary way."[32] But as Irene Spry has pointed out, these estimates omit the cost of labour.[33]

Labour was crucial in establishing a farm, and the cost of hiring it could easily spell the difference between success and failure. But immigration promoters tended to emphasize how much farmers could earn on the farms of others rather than how much they would have to pay in wages. They assumed that the most important non-cash resource settlers possessed was their capacity for hard work, and they expected settlers to use their energy wisely in building up their own farms and in earning extra cash on the farms of others. Although data are not available to trace the numbers of homesteaders whose failure to prove up can be attributed to working on the farms of others, the volume of applications for extensions of the homesteading time limit attests to the difficulty farmers faced in weighing the delicate balance of time and labour expended on their own farms against the opportunities to earn cash elsewhere.

Still, the promise of independence through farm labour was a powerful draw. Books, pamphlets, and reports echoed the advice of Alfred Pegler, a reporter for the *Hampshire Independent* of Southhampton, who undertook an investigative visit to Canada in 1884. He joined other voices of the day in urging men to come to Manitoba on the first of May with 100 pounds in their pockets. For this sum the newcomer could build a house, buy a yoke of oxen, and purchase tools and seeds to carry him through until he sold his first crops. After breaking twenty-five acres for the first year's crop, he could earn enough cash for his winter's provisions by hiring himself and his oxen out to his neighbours at $4.50 to $5.00 per day. In his second year, he could crop the original twenty-five acres and break another twenty-five.[34] Within one year the newcomer could count on an income that came from the work on his own land as well as on that of others.

Both types of earnings were necessary. The tactic of the successful Dubliner and countless others like him – to homestead and to work for wages concurrently – was recommended to all but the most financially secure settlers. The Canadian Pacific Railway assured intending immigrants that wage work for other farmers was readily available and need not interfere with operations on the home farm. "The settler opening a new farm can always find plenty of work among his neighbors," declared a Canadian Pacific Railway advertisement for the North-West Territories, "after he has done his own breaking and back-setting and cut his hay."[35] It was the juxtaposition of these opposite forms of income that would lead both to the development of western agriculture and to independence of the individual farmers.

For men with no capital at all, advocates of western settlement had other advice. In order to earn the necessary stake to begin farming, ambitious and hard-working men could readily find work in any number of Canada's developing industries. Reports like Pegler's were optimistic, but were tempered by a recognition of the uncertain labour market. Pegler warned that demand for labour outside agriculture was unsteady, and that although highly skilled workers could earn as much as $3.50 to $4.00 per day at piecework, the "poorest class of labourers" could expect only $1.50.[36] At the time of his report, he noted that there was no demand for labourers, carpenters, or bricklayers in Toronto, and he reported that the Canadian Pacific Railway, while calling for 2,300 men in the North-West, had just reduced its work force by 10 per cent.[37]

Agricultural labour was presented as a more certain route to the savings necessary for successful farming. Even in 1876, when Canadian workers were suffering from such high unemployment that their plight drew attention in the press, the Minister of Agriculture endorsed the blithe assertion of Charles Foy of the Belfast Agency, who reported that "I am sure that there are no good ploughman nor farm labourers amongst them." Foy deplored the "gloomy reports" and called on newspapers to exercise logic to see their error: "by adducing the simple fact that every genuine farm labourer could get immediate employment, and that if the calling of every one of the thousands out of employment were given, they would see that there was not a single genuine farm labourer amongst them."[38]

Pegler, the English journalist, shared the popular and widespread belief that western Canada was a land of unlimited agricultural potential. Success stories received wide publicity. Notable cases were those of Christian Troyer of Alameda, NWT, who explained that "I borrowed 8 [pounds] to come here with" and claimed to have a farm worth 2,000 pounds, and Henry Proctor of Woodlands, Manitoba, who arrived with "Nothing" and was now worth 2,400 pounds.[39] Apparently, men could start out with little more than a filing fee and a strong will. Even intending settlers with no capital at all faced excellent prospects. Szikora Mihaly of Kaposvar, Assiniboia, had emigrated in 1888 "with a debt of $70," but had soon parlayed hard work into "320 acres of fine land ... three draught horses, six head of horned cattle, and pigs and poultry enough for our wants."[40] W.E. Cooley of Birtle, Manitoba, recounted that when he had arrived in 1887, his "earthly possessions . . . were $1.75, a wife and several children." Through a combination of waged farm work and tenancy, he was soon "the owner of 800 acres, and I think as fine a residence as there is in the district."[41] For "the large numbers whose only inheritance is their brawny arms and determined will,"[42] farm work was readily available, and men like the Dubliner were offered as proof.

The widely publicized need for farm labour was no exaggeration, and the problem was evident even to observers outside the farming community. Distressing results of shortages drew comment from the North-West Mounted Police in their annual reports. "As a rule, too little fall ploughing is done in the North-West, and there is consequently too much hurry among farmers in the spring, and large tracts of land are sown but not sufficiently worked," observed Commissioner Lawrence Herchmer in 1886. "Nearly all the farmers work too much land for their own strength."[43]

Farmers besieged government agencies with requests for hired help. J.S. Armitage, agent for the Manitoba Department of Agriculture and Immigration in 1887, left a file of letters outlining his difficulties. "Men for farm labour are scarce," he explained to a Salterville farmer, "and unless more men come in I am afraid I will not be able to get you one."[44] The farmers' desperation for labour only compounded the problem of co-ordinating supply and demand, as Armitage complained to another farmer: "Your application for a farm labourer is received and I fully expected one to have gone out to you yesterday, but men are very scarce and we cannot always depend upon them keeping their word, as some man may come in and take them off."[45]

Private agencies whose interests coincided with government policy added their voices to the call for agricultural labourers. The Canadian Pacific Railway, the largest private landholder in the West and one of the many organizations actively involved in placing farm help, was one of the most vocal advocates of farm labour as a prelude to farm ownership. It encouraged men to come to the West, where farm work was plentiful, and called on local farmers who were potential employers to back up its claims. Testimonial letters from farmers in search of hired help were extravagant in their assessment of the opportunities for farm work. When James Kelly, a farmer at Arnaud, Manitoba, wrote to the Manitoba government agent in Liverpool in 1891, his letter subsequently appeared in a Canadian Pacific Railway pamphlet. "A thousand farm laborers would get work here at present," he declared, "at wages from $50 to $60 and board per month."[46] By 1894, observers of the prairie West were assuring their readers they could provide "hundreds of thousands of instances" of opportunities for farm workers to "live well and get good wages."[47]

Good wages meant more than simply rates that were competitive with those in other industries. Wages represented land. Men were eager to know just how quickly the payment for their labour could be turned into a working farm. If a man were to believe that he could earn the promised $50 or $60 a month plus board, he would be able to save the recommended $500 in less than a year. Even the most exuberant literature did not hold out such a definite guarantee, but information for settlers was filled with equations for translating labour into land. Calculations varied, but all were confident. "Practical workers, if steady reliable men, could do well here," declared Colonel P.G.B. Lake of Grenfell, Assiniboia, in a Canadian Pacific Railway pamphlet. "Good wages are paid, and an honest man starting with no capital and saving carefully from his wages, should find himself fairly upon his feet on his own account, in at most five years."[48]

The message to intending settlers was clear. In other industries, the labour force remained the labour force, but in the agricultural industry of the prairie West, men were encouraged to enter waged labour for the express purpose of leaving it. Immigration pamphlets held out the promise:

> The laborer is happy and contented; he is only waiting for an opportunity to get a farm of his own and become as independent as his employer. With a farm free from debt; his fields of ripening grain ready for harvest; with herds of cattle on his pasture lands, and flocks of sheep feeding on the hillside; dairy and poultry providing the household with groceries and many other comforts; schools for his children in the immediate neighbourhood; churches close at hand, and such other social advantages as he desires within easy reach – what more is required for a happy existence?[49]

The Lure of Free Land

The response of ambitious newcomers was equally clear. For Herbert and Dick Church, who moved to the North-West Territories from London, England, in 1886, land was "a means of getting a permanent and independent livelihood, and perhaps something more," and hard work was a fair price to pay.[50] No other industry could offer such a direct translation of wages into capital into independence. To intending settlers, the promise of land in return for their labour was the cultural condition of their decision to enter prairie agriculture; the ability to exchange their labour for land was the structural condition shaping their actions within the industry. Their hard work, perseverance, and independence would be repaid with landownership, which, they believed, would directly enhance their material and social conditions.

The common denominator was a search for improvement. Men who came to the West were seldom those who were leaving absolute impoverishment at home. Rather, they were ambitious, individualistic men whose strivings were channelled into a desire for land they could own, live on, and farm. They constituted a stratum of migrants who were anxious to hold on to what material and social gains they had made and who hoped to build on their strengths to create an even better economic and social position for themselves as prairie farmers.[51] The specific criteria for what constituted improvement were as varied as the social and economic backgrounds and national origins of the men who came.

Although Canada actively recruited immigrants in many countries, the greatest number of prairie settlers came from within the Dominion. Men from Ontario particularly flocked to the West. As early as the 1850s they began to regard the West as their patrimony and to press for annexation of the western prairies. This is hardly surprising. Agriculture in most of Ontario had passed the pioneer stage of inexpensive and accessible farming, if such a stage actually ever existed.[52] David Gagan's study of Peel County shows an "apparently stagnant community" by the 1860s. Nor was Peel an isolated case, for "demographic decline or stagnation was prevalent among the south-central and eastern lakeshore counties of Canada West."[53] The limits of agricultural expansion had been reached; further growth could only be achieved through consolidation and more intensive use of the land. Farmers in Peel enlarged their holdings, usually at the expense of smaller, less successful farmers, and altered demographic patterns of marriage and inheritance. Typically, a farm was bequeathed to one son, who was then held responsible for his brothers and sisters. Marriages and the arrival of children were delayed. These patterns relieved the pressure on land, but created social pressures on the people involved. Farmers' sons who did not inherit the land, as well as immigrants and newcomers, were effectively excluded from establishing farms. When prairie lands became available in the late 1870s, they moved west.[54]

Their expectations were high. Men who based their decision to move on the widely disseminated settlement propaganda felt assured of independence and prosperity, a far cry from the limited horizons of a lifetime of poorly paid agricultural work at home. Men who preceded them to the West underscored the promise. J. Downie of Oak River, Manitoba, maintained that he was "perfectly satisfied, and would not go back to Ontario to farm if paid for it."[55] The feeling was widespread. George Wood, a self-proclaimed leader of the community of Birtle, Manitoba, declared that he did not know "a single instance of a sober, industrious person who has not benefitted by coming here, and I do know of many who always lived 'from hand to mouth' in Ontario, who are getting rich."[56]

The next major group to be drawn to the West was immigrants from Great Britain. They represented a wide variety of backgrounds, from farm labourers to the sons of the well-to-do. For some it was an opportunity for adventure, but for most it was a chance to improve their material position. Men who had been in business came looking for a change. Men who saw a threat to their economic and social position came looking for a chance to secure their future. Men who had been unable to establish themselves came looking for new opportunities. Men who saw their families living in want came looking for a new life.

Farm labourers fell into the last category. Throughout the latter part of the nineteenth century, large segments of the British agricultural work force were becoming increasingly impoverished. Their efforts to improve their position resulted in the formation of a National Agricultural Labourers' Union.[57] The Union fought for better working and living conditions at home and was instrumental in raising the wages of farm labourers. But prices were rising, too. The major problem was the glut of farm workers on the labour market. The Union's solution was to encourage and financially assist the removal of superfluous workers.[58] In its efforts, it was part of a larger and concerted attempt by reform agencies and the government to remove Britain's surplus labouring population. This dovetailed neatly with Canada's own desire to attract immigrants who would provide the labour necessary for agricultural improvement.

An agreement in 1874 with Joseph Arch, the president of the National Agricultural Labourers' Union, was the first concerted attempt to resettle farm workers in Ontario, where they might continue to find employment. The agreement was designed to provide a labour force that was inexpensive, hard-working, and unambitious. Timothy Demetrioff's study of the agreement indicates the scarcity of such qualities among the small number of Canadian men who were willing to undertake waged farm labour.[59] By contrast, the English agricultural labour force was "ready to work at anything by which they can honestly earn their bread," according to the *Ottawa Times*, "and this at a very low rate of wages, indeed, and without any prospect of improvement."[60]

In 1875, the Dominion government stepped up efforts to attract farm workers and debated the merits of assisted passage. British agents described the "large numbers of people" who would leave at once "if we could provide the means to pay their passage, which they were unable to do themselves."[61] An agreement with steamship lines allowed considerably reduced rates to farm labourers and their families. Other attractions were improved standards of living and amenable social conditions. Following the agreement with Joseph Arch and the National Agricultural Labourers' Union, a rival union visited Canada hoping to gain similar concessions. In 1875 a delegate of the Federal Union of Agricultural and General Labourers returned a glowing report:

I may say that throughout my tour, I everywhere found the diet and treatment of the labourer, when boarded by his employer, far superior to what he could obtain in England. Animal food at every meal is the usual rule, and working as they do, side by side, there is not, as in older countries, a wide gulf of separation between employers and employed.[62]

Canada was anxious to receive British farm workers. "It has been found, almost without exception," declared the Minister of Agriculture, "that English agricultural labourers with families, [are] the class most desirable to bring to Canada."[63]

But farm labourers faced a bleak prospect. Wages and conditions deteriorated when Ontario agriculture entered a slump that made it difficult for farmers to pay even low wages. The British example of union success among farm labourers spurred a similar effort in Ontario. A treatise on *Farm Life As It Should Be and Farm Labourers' and Servant Girls' Grievances* appeared in the 1880s, proposing a union for farm labourers in Ontario. The booklet extolled the virtues of agriculture, but deplored the conditions into which farm labour had sunk. "Tilling the soil should be man's noblest work," wrote Edward Amey, "but unfortunately avaricious and short-sighted people have made it slavish and degrading."[64]

Waged farm labour was extremely unattractive. Farm children who had expected to become farmers themselves were dismayed by the prospect of working permanently on the farms of others. At the same time, the physical toil of farm work was losing its social acceptability. Farm consolidation and post-pioneering prosperity meant that the arduous labour of farm work was no longer the social leveller it had once been. Alison Prentice has shown that with the increasing mid-century emphasis on education and gentility, both for children and in its technical application in economic pursuits, there was "a strong suggestion that the physical aspects of farming were really degrading."[65] Farming was still a noble profession, but farmers were expected to take advantage of education to distance themselves from the manual labour. The corollary was that "there was little apparent hope for the mere agricultural labourer or farm servant."[66]

Men who moved to Ontario to begin farming soon found that farm labour held no prospects and no immediate compensations. George Tuxford left his family firm in North Wales in the 1880s to try his hand at farming. He began his new career in Ontario and worked on three or four farms before moving to Moose Jaw to take up his own. His experience led him to advise his parents to urge his younger brother to come directly to the West, and not to stop in Ontario as he had done. "In one way if he worked there for a year, it might lessen his liking for farming considerably," he wrote to his parents, "for a hired man's work there is very different from England. It is simply slavish."[67] Amey's proposed union might have been a vehicle for improvement of the conditions, if not the prospects, of farm labour, but it either never materialized or has left no accessible record. Ambitious men simply moved on.

Canadian efforts to attract men near the bottom of the economic scale were soon redirected. The agreement with Joseph Arch's union was short-lived. Ontario's economic slump caused a reaction against the distribution of public largesse to men who would enter an insecure labour market. At the same time, improved wages and conditions for farm workers in England caused them to lose interest in moving to the colonies simply for better employment opportunities.

These conditions had the opposite effect on a group of men with the agricultural experience that Canada was seeking. Tenant farmers began to fear they might lose what economic and social status they had gained, and they began to consider emigration as a likely route to secure their position and even to improve it.[68] Readers of immigration literature were encouraged to ask themselves: "Can I hope to live there with greater comfort and less anxiety for the future of myself and my children than in the old country?"[69] These men had capital. "I am certain that this tenant farmer class will be of benefit as they had some means with them," reported Canada's immigration agent in Liverpool, since "they are not as entirely dependent on their manual labour for the support of themselves as the agricultural labourers."[70] Canada began to woo them.

Farm ownership was expected to have the greatest appeal. Although Canada sought settlers with enough cash to establish themselves quickly and without having to draw on Canadian resources, the country also needed men with limited means who would provide the labour for a rapidly growing industry. Tenant farmers filled the bill. Farm labourers were still actively recruited, but attention was increasingly directed toward men with modest means who could be counted on to provide the labour not only for their own farms but for those of their neighbours. The draw was economic and social improvement:

> It is for the possibility it affords of elevating himself, and above all his children, in the social scale, and not from any mere increase in wages, that Canada is to be recommended as a home for the British labourer. A man with so little ambition as to have no hope or wish to be anything in the future but a labourer, and who only desired increased wages as a larger fund for self-gratification had better remain at home.[71]

Only in western Canada were prospects for such improvement still widely available. Ontario was the point of arrival for great numbers of British immigrants, but for those with aspirations to independence it was a mere stopping place to earn some cash and to learn rudimentary farm skills. The most ambitious soon joined the Ontarians and moved west.

There they were joined by settlers from the prairie and midwestern United States who faced slender prospects at home.[72] The increasing difficulty American farmers' sons and hired hands faced in achieving farm ownership was revealed as early as 1880, when the American census showed an alarming extent of tenancy. In the north central states tenants operated more than 20 per cent of the farms. During the following decade the rate of tenancy in the westernmost of these states nearly doubled, and it came close to doing so again in the 1890s.[73] Commentators at the time credited the phenomenon to a surge of men taking a step up the agricultural ladder to farm ownership, although later analysis concluded that, in fact, it represented a shrinking of agricultural opportunities.[74] At the same time, farm consolidation, mechanization, specialization, and falling agricultural prices conspired to reduce significantly the economic and social position of waged agricultural labourers.[75]

Men who sought farm ownership found that their position on the bottom rung of the agricultural ladder was likely to be permanent, as the number of men who became farmers after stints as tenants or labourers began to dwindle after 1880.[76] The opportunity to realize their ambitions by moving further west was also coming to a close. In 1890 the last great land sale in the western territory took place. Sub-humid lands were still available under the Kincaid Act, but these were marginal. As Gilbert Fite has indicated, "good opportunities for new settlers on the public domain were gone."[77] Men began to look northward.

Continental Europe was the other major contributor to western settlement. Some of the migrants from the United States were of recent European origin and saw more fertile fields north of the border, but the majority of European immigrants came directly from Europe, sharing with other newcomers a common goal of farm ownership. Throughout Europe, as in Great Britain, it was almost impossible for any but the very wealthy to obtain land except through tenancy. Agricultural labourers could not even hope for this much.[78]

Although the prime attraction of North America was free or cheap land, even an immediate improvement in the conditions of agricultural labour could act as a draw. The appeal of independence was very strong for farm workers who were accustomed to strict controls. Lars Peter Erickson, a Danish farm worker who came to Canada in the 1880s, brought his *Skudsmaalsbog*, a booklet recording all his agricultural employment. Erickson explained that every agricultural labourer was required to carry such a booklet and to have it signed by each employer.

"It was impossible to move about the country," he noted, "unless you had one of these." The possibility of farm ownership was Erickson's prime motive in coming to Canada, but even in the short term he looked forward to agricultural employment free from "this regimentation." He also looked forward to better working conditions, complaining about the "poor treatment of workers" back in Denmark. At his last job there, Erickson's only time off had been every second Sunday, and that was "the best place he had worked at."[79]

Men like Erickson fully expected to put in some time as wage labourers on the farms of others, but their ultimate objective was farm ownership. The attraction was so strong that the numbers and proportion of Europeans moving to the West increased steadily. In 1881 the foreign-born population, composed primarily of Europeans, made up less than 14 per cent of the prairie population. A decade later it had grown to more than 18 per cent, and by the beginning of the new century the foreign-born had surpassed the British-born segment of the population by rising to almost 25 per cent.[80] Like the newcomers from Ontario, Great Britain, and the United States, Europeans came to the West because they wanted to farm but had no hope of farm ownership in their homelands.

In the closing decades of the nineteenth century western Canada began to gather a population with a desire for self-improvement, with ambitions extending to the economic independence they believed would result from farm ownership. At the same time this conviction limited them. The prairie West drew men, whatever their origin or ultimate plans, by the promise of farm ownership, to be achieved by their own individual labour. Their future was assured in glowing terms: "for the industrious, self-reliant, frugal and observant man, who, while preparing to do his duty by his present employer, looks forward to owning his own acres and securing still brighter prospects for his offspring, Canada affords chances inferior to those of no other land."[81]

The concerted efforts of governments and immigration agents to woo agriculturalists, and to instill in them expectations of plentiful farm work, resulted from the disappointingly slow rate of agricultural growth at the end of the nineteenth century. Since this growth was being fuelled primarily through acreage expansion, which was being carried on by labour rather than the technology that capital could buy, it was clear that it could not continue without an increase in the labour force. And when the labour force did multiply in the opening years of the twentieth century, agriculture expanded enormously. Economic realities thus played a large part in bringing a labour force to the prairie West and in keeping it there. Canada continued to be the "Last Best West,"[82] offering land for homestead and sale long after entry into farm ownership had become a vain hope in the more developed agricultural regions.

Immigration literature promised land in exchange for hard work.
(National Archives of Canada, #C83565.)

Illusion was an even more powerful force. Although the prairies have left ample records of farm dreams that were dried out, or hailed out, or foreclosed upon, there were always enough tales of settlers who did secure their own farms and attain their own standard of success to validate the claim that it was possible. Even the failures served to reinforce the belief that individual effort and hard work would be rewarded. Those who abandoned their attempts at farming were characterized as unadaptable or unwilling or unprepared for the difficult early years. Their absence made self-defence impossible, and they appeared as shirkers who had been weeded out, leaving behind those who were willing to continue the struggle. Those who remained were often forced to lower their expectations for material improvement and economic independence. As Lyle Dick has suggested about long-term farming progress in Saskatchewan, "persistence was not necessarily an advantage."[83] Yet long-time farmers looked back with pride on their hardships, counting the farm deed in their hands as the ultimate success.

Even when land became scarcer and labour became more plentiful, immigration and settlement propaganda continued. It did more than merely provide a labour force – it legitimized the structural and cultural conditions necessary to create a labour supply that was both cheap and abundant. Individual farmers may have been cash-poor, but the industry as a whole relied on the federal government to underwrite its labour costs by providing land as a wage supplement. By the time this became less feasible in the post-war period and beyond, the pattern of expectation had been established and rationalized. Farmers continued to depend on low-wage labour. Men continued to work for meagre wages in the anticipation of a future reward in land.

Both the prescriptive and the proscriptive nature of the promise had ensured that those who answered its call fit the description of the typical self-made man described by Allen Smith. The pioneer farmer, and the man who aspired to become one even long after the pioneer era closed, was "free of all constraint and interference, quite literally able to shape his world as he wished." And this was to be achieved entirely through his own effort. His "abundant and fulfilling future" was to be attained by "no more nor less than his own capability."[84]

PART II

EXPANSION, 1900-1918

4

Agricultural Labour
as Apprenticeship

"I won't be long in having a place of my own." [1]

In the early years of the twentieth century, the promise of *A Farm for Two Pounds* drew Harry Baldwin to the prairie West.[2] Baldwin achieved his dream by toiling his way up from farm labourer to farm owner, then published an account of his accomplishment. His story is romantic and heroic, similar to that of dozens of others left by prairie pioneers who wrote themselves into the saga of the developing West. His view of himself and his place in the new economy and society is representative of men who entered the shifting world of labour-capital relations during the early years of the wheat boom. It was a time when the agricultural industry was rapidly expanding and when the men who laboured in it were negotiating their roles to meet their expectations.

"That farm of my own was the lure," Baldwin recalled, explaining his flight from the English Midlands. "I craved an outdoor life, but in a land so congested as ours, however a man toiled he could never become anything more than a farm labourer." Baldwin's ambitions were modest, but they filled his horizon: "In the west my own fields, the shocks of my own grain, awaited me."

Baldwin embarked on his agricultural career as a hired hand in Ontario, where he hoped to learn Canadian methods. He was proud of his ability to milk cows and drive a team of horses, but he had much to learn. "Although the principles of the farming game were familiar," he remarked, "being new to the ways of the country, I naturally made mistakes." Baldwin regarded his education as a necessary prelude to farm ownership, and although he chafed about "the fetish of the Ontario farmer – work for work's sake," he relished other aspects of farm life. "How good the food was!" he recalled. "My meals were a joy."

Baldwin was also receiving lessons in labour-capital relations. He found his work "dull, weary, exhausting toil" and fixed the blame on his employer, whom he accused of "slavery and slave-driving." His boss returned the compliment by chiding Baldwin for his short stature and slight build, casting doubt on his likely productivity. Baldwin capitulated and took a cut in pay. But the experience stiffened his resolve. With a "fixed determination to start on the road that led to a farm of my own," he found work on another farm. His second experience could not have provided a greater contrast. Again, Baldwin understood the distinction in terms of individual differences. He credited his employer with creating an atmosphere of equality, "no master and man stuff about it." His year on the farm was rewarding. "I learnt quickly under Rube's direction," he explained, and proudly reported that "no better relations could have existed between a hired man and his employer."

But Baldwin was anxious to enter the world of farm ownership, and "there was a glamour about the very name of the West that struck an answering chord in my romantic young heart." As he joined the harvest excursionists, he saw himself among "the bronzed young Apollos carrying sheaves of purest gold over stubble as golden under an azure sky." The thrill of such glamour seemed payment enough, but Baldwin also anticipated earnings of such princely proportions that he need only labour a "short spell and then – my horses, my sheaves, my homestead!"

Once again reality pulled Baldwin back to earth. "Life was a dreary misery on that Manitoba farm," he recalled, and stooking was the "most monotonous job in the whole gamut of labour." Yet as an agricultural labourer, he was developing qualities that would stand him in good stead as a farm hand and eventually as a farmer. He learned resiliency and began to joke that "the monotony was relieved by constant fault-finding." He took pride in his work and bragged that his abilities were recognized when his sheaves weathered a storm as well as any others. He cultivated feistiness and related how he settled a dispute with his co-worker: "I smote him right on his dapper little moustache." By the time his work on the farm began to pall and his boss began to criticize him, Baldwin had no qualms about quitting. When the boss refused to pay his wages, the two came to blows. Baldwin lost the fight but won his wages. "Very shortly after I was on the road once more," he recalled, "with a patched eye, a bruised face and a big lump on my head, but with quite a few dollars in my pocket."

Despite the arduous work and often unsatisfactory conditions, Baldwin remained enamoured of prairie agricultural life. Learning to operate complex farm machinery filled him with pride. Catching the "faint odour of gasoline and the reek of sweat and harness" caused his heart to catch with a "passionate love of the west." His resolve to make this his life's work stiffened, and after a brief stint in a lumber camp, he returned to waged agricultural labour.

More cautious now, Baldwin recognized the economic and social particularities of prairie agriculture and determined to wrest what he needed out of the bargain for his labour. He weighed the relative benefits of higher wages against likelihood of collection, of hard work against generous room and board, and based his decision on his own sense of ability to control his conditions and handle his employer. "Shucks, I liked the old skeezicks," he declared of his new boss, "and I wasn't such a kid, now. If he came any funny stuff I'd find a rock to heave at him."[3]

Baldwin's education as a hired hand was complete. He had not only learned how to farm the Canadian prairies but had come to understand the particular relationship between labour and capital. The lure of independent farm ownership had drawn him into agricultural labour, and his experiences of that labour and of the relationships within which he operated had fostered this sense of independence. Baldwin perfected strategies to deal with his conditions of work, and with his employers, that were highly individualistic and ideally suited to the conditions of prairie agriculture during the period of settlement and agricultural expansion.

The early years of the wheat boom were the heyday of the agricultural labourer because the shortage of labour and the plenitude of land coincided to allow easy access to both farm work and farm ownership. But the opportunity was short-lived, curtailed by the post-boom depression and the First World War. Nor was the quest for independence without difficulty and complexity. Men found themselves in an unconventional position that was neither clearly labour nor clearly capital. The strategies they developed for achieving their various goals reflected this anomaly: they acted both as aspiring farm owners and as active members of the working class.

Agricultural Expansion: 1900-1918

Harry Baldwin came to the West at the beginning of the wheat boom, a period when the production and export of wheat brought long-awaited prosperity. In the first two decades of the twentieth century, the number of farms increased nearly fivefold and improved acreage grew eightfold.[4] The most striking increase was in wheat production. In 1890, the prairies produced 16.5 million bushels of wheat. Twenty-five years later a bumper crop measured in at 360 million bushels.[5]

The boom was based on expansion. Ready markets and high prices for wheat drew agricultural settlers bringing capital and labour. The promise of agricultural growth attracted developmental and investment capital from government and private investors. The combination of a rapid and massive increase in labour's productive capacity and capital's lavish investment created a striking but temporary phenomenon of great prosperity, providing a heady atmosphere of limitless potential.

As the nineteenth century drew to a close, the trickle of newcomers to the West turned to a flood. Homestead entries soared to a peak in 1911,[6] translating into a rapid increase in agricultural settlement. The number of men engaged in agriculture increased from close to 43,000 in 1891 to almost 279,000 by 1911.[7] During the first decade of the century alone, the rural male population nearly trebled.[8]

Expansion was also made possible through technological breakthroughs, even when this was not the intended effect. Seed selection is one example. Although Red Fife wheat had ensured a world market for Canadian wheat, it often did not ripen in the short prairie growing season. The search for a faster ripening strain was rewarded with the discovery of Marquis wheat, a hybrid variety introduced in the West in 1907 and widely distributed by 1911. Marquis matured eight days earlier than Red Fife, could withstand dry conditions better, and yielded, on average, seven more bushels per acre. The greater security provided by a shorter ripening time and a heavier yield acted as a further encouragement to wheat growers to expand their acreage, and the shorter growing season and drought resistance enabled them to push the agricultural frontier further north into areas with a shorter season and further south and west into slightly drier areas.

Other technological innovations were aimed at improving grain yields. Dry-land farming techniques, well established by the turn of the century, were refined. Successful practitioners of dry-land farming developed and advocated techniques for improving the quality of seed, for controlling plant diseases, and for careful cultivation and tillage practices that would preserve moisture and eradicate weeds. John Bracken and Seager Wheeler were two successful prairie wheat farmers who joined their voices to the agricultural journals and experimental farm experts to disseminate information on the new techniques.[9] They advocated intensive tillage, with extra harrowing and packing, over the much faster but "permanently wasteful" practice of burning stubble. But these practices were labour-intensive. Other technological improvements aimed at increasing the efficiency of labour. Modifications and improvements in implement design and the introduction of new machinery often led to greater agricultural productivity per man-hour, although these changes did little to reduce overall labour requirements, since the time saved was used to produce larger crops.

Rapid settlement and technological innovation translated into rapid development, but it also meant that the limits to expansion were reached very quickly. Manitoba was the first to fill up. By 1903, there was little land left for homesteading, and even purchased land was rapidly taken. By 1908, good land close to railways was gone. Farther west, most of the best land between the parklands and the dry area of Palliser's Triangle was taken by 1908.

A sense of urgency underscored the drive for expansion. The demand for land was putting settlers into areas in southern Alberta and southwestern Saskatchewan that could only be made productive through costly irrigation projects. The costs of agricultural over-expansion were soon evident. Drought brought homesteaders to the edge of starvation.[10] Over-expansion had broader results as well. The rise in production stalled in 1911 and grain prices fell markedly in 1912.[11] The downward trend continued the next year, with no relief in sight. The entire prairie economy was affected. Investment capital began to dry up, railway construction slowed in 1912, urban building starts fell in 1913 and 1914, and unemployment became more than a seasonal phenomenon.[12] The prairie economy was reeling.

War revived the agricultural industry. The first wheat crop harvested after the declaration of war was by far the largest on record, and it found a ready market in war-struck England and Europe. Prairie wheat production rose to more than 360 million bushels in 1915, from an average of almost 185 million bushels during the previous three years. Every bushel that could be produced was snapped up at once.

Prices rose dramatically during the war. From $.89 per bushel in 1912 and 1913, the Lakehead price of No.1 Northern wheat jumped to $1.32 in 1914, reached $2.05 in 1916, and was finally capped by the federal government in 1917 at $2.21.[13] The farm value of the 1915 wheat crop was $324.8 million, approaching three times the 1914 figure of $135.9 million, and the farm value of the 1915 harvest of all major crops doubled to $441.3 million from $220.9 million the previous year.[14] By 1917, the farm value of major prairie crops had risen to almost $630 million.[15]

Commercial farmers reacted quickly to expand their holdings, even when this meant stretching their credit to the limit or moving into areas considered only marginally arable. The decade that encompassed the war witnessed an increase in the number of farms of more than 25 per cent and in total acreage of more than 50 per cent. Improved acreage almost doubled.[16] Prairie wheat acreage began a steady rise in 1915, increasing to 13.9 million acres from the 10 million acres of 1914. By 1918 it was up to 16.1 million acres.[17]

But larger acreage of wheat required more labour, while enlistment and wartime job opportunities elsewhere drew thousands of men from prairie farms. This was particularly evident when it came to harvesting the wheat that held such promise to the fortunes of prairie farmers. The number of harvest excursionists from outside the prairies, which had grown from an annual average of under 4,000 in the last decade of the nineteenth century to an average of close to 18,000 in the first decade of the twentieth, shot up to more than 42,000 in 1917.[18] The pressing need for labour finally pushed agricultural technology in the direction of reducing labour needs rather than simply making labour more productive.

The gamble of expansion paid off in a doubling of wheat exports during the war. By war's end wheat had become the largest single Canadian export in terms of dollar value.[19] Canada had become the second largest world exporter of wheat, averaging almost 165 million bushels annually during 1915-17, with 90 per cent of this from the western provinces.[20]

The economic roller-coaster of boom and bust that characterized the agricultural industry during the first two decades of the twentieth century was reflected in the relations between labour and capital. Farmers juggled their drive for expansion with their concomitant increase in labour needs, attempting to bring more land under cultivation while gambling that their labour needs would be met. The rush of settlers to the West offered no solution. As long as land was available and affordable, men placed ownership at the top of their list of priorities, entering waged agricultural labour only when and if it promoted this goal. Labour shortages thus continued to plague the industry, although the absolute labour shortages of the last decades of the nineteenth century became instead relative. It was a circular process – as men rushed to take up farms of their own, they increased overall labour needs.

The combination of agricultural expansion and the use of technology to make expansion possible added up to overall agricultural prosperity. In the first decade of the twentieth century and during the First World War, the results of this prosperity generally filtered down to individual farmers. Even so, distribution of the agricultural bounty was very uneven. Years of spectacular crops were followed by years of equally spectacular failures, and for individual farmers the result could be fortune or insolvency. In periods of economic setback, farmers suffered even more. The industry as a whole could easily absorb the annual fluctuations and even the more serious and long-term slumps, but individual farmers could not.

Their response was to tackle the problem head on. Farmers lobbied the federal government to reshape tariff policy and transportation rates, worked with provincial governments to ensure controlled and adequate storage facilities, and organized themselves to enter the field of marketing.[21] Through these ventures, prairie farmers used their combined voice to demand protection and promotion of their interests. By 1916, Henry Wise Wood, president of the United Farmers of Alberta, was able to articulate a feeling of unity and direction among farmers. "We are a class organization, it is true, but we are the basic class," he declared. "We represent a rising of the people, the great common people, en masse, in an upward struggle."[22]

It may have been economically and politically useful for farmers to see themselves as a single class, although recent scholarship has done much to uncover evidence of an increasing degree of stratification as farming communities moved beyond pioneer conditions.[23] But in terms of the relations between labour and capital, farmers' awareness of their special status and their actions to overcome their disadvantages gave them very little edge when it came to controlling their own hired hands.

Farm Workers as Apprentice Farmers

Hired hands did not intend to remain hired hands. The economic boom that brought more settlement and greater production meant expanding opportunities for farm employment and for eventual farm ownership. But these were conflicting aims, and farm workers who aspired to ownership became enmeshed in the contradiction. The inconsistency of their class position was reflected in their responses to the situation and their methods of dealing with it. Although hired hands put most of their energy into strategies for removing themselves from the working class, they also pursued strategies to improve their position as members of the working class. Few men took the time to ponder the dichotomy as they pursued their primary objective of farm ownership.

Newcomers were encouraged to bring with them enough capital to begin farming at once, but many came with the intention of beginning as labour and climbing the agricultural ladder to ownership. The "agricultural ladder" as it existed in the prairie West deserves attention for its effect on labour-capital relations. Its importance lies as much in what it promised as in what it delivered. The agricultural ladder on which a hard-working and frugal young man could climb from farm labourer to farm tenant to farm owner was a notion familiar to North Americans.[24] The ladder did exist, but it was a fleeting phenomenon, operating for short periods only when land was plentiful and labour and capital were in short supply. It was limited to the frontier, existing in southern Ontario during the early part of the 1800s to about mid-century, and in the United States moving westward with agricultural settlement until about the turn of the century.[25]

But by 1900 even the western United States could no longer provide access to agricultural independence. In Oklahoma more than 43 per cent of farmers leased their land, and the pattern was similar in Texas and Kansas. In the states of the western Middle West, tenancy was increasing, not as a step up toward ownership but as a step down. The alarm caused by the increased tenancy revealed in the 1890 census seemed well founded when the 1900 census showed an even greater increase.[26] Frederick Jackson Turner's startling thesis of 1893 that the frontier, which had provided "the promotion of democracy," "economic power," "political power," and "intellectual traits of profound importance," was closed, seemed accurate.[27] In an era of agricultural consolidation and mechanization, small farmers were driven out of business or into tenancy.

Farm workers were unable even to achieve tenancy. The increase in agricultural income that followed the 1896 economic recovery was very unevenly distributed, with waged farm workers receiving less than half their share of the returns. In Iowa, for example, farm labourers made up 35 per cent of the agricultural population, yet received only 6 per cent of the gross returns to agriculture.[28] The north central states sent the Canadian prairies the greatest number of American settlers.

Fred Pringle was typical of young Americans who knew they could not climb the agricultural ladder at home. In 1909 he moved north from Montana with a clear intention to make his agricultural future in Canada. His careful record of expenditures suggests he had a nest egg of no more than $100. In the autumn before he came, he had hayed, threshed, and husked corn on several farms, then turned his hand "to work at the Barber trade" for a few months.[29]

When he "Pulled out of Billings" on April 22, 1909, Pringle spent $13.90 for a railway ticket through Great Falls and Sweetgrass to Lethbridge, then another $3.15 to reach Calgary and finally Stettler. He was in no great rush to reach his new home, for he stopped in Lethbridge to attend the Wesley Methodist Church, write a letter, and admire the "big bridge 507 ft high." In Calgary he "loafed around" for two days, waiting for his trunk and writing letters. From Stettler he travelled for a day to Lacombe, where he stopped and "Danced 3 times," spent a dollar on whisky, sent postcards to the folks back home, and "went to church twice." He then returned to Stettler, where he picked up a little cash by "Slinging ties all day" for $1.50. Finally, on May 7, he "Filed on a Canada claim" and sent for his trunk. Despite the terseness of the entry, there is a note of tempered pride. Pringle at first wrote in his diary of "<u>my</u> Canada claim," proudly underlining "my." He was bitten with caution, however, and crossed out the "<u>my</u>" to replace it with the neutral "a."[30]

With only $41 left in his pocket, Pringle began his homesteading venture by seeking to accumulate cash. He found work the next day "Building fence for Mr. Hargreaves," and three days later he "Hired to Geo. Hoskins for 1 mo. for $35." Pringle spent his first summer and fall on the Canadian prairies working at different farms. On September 9, he "Finished month['s] work for Wisler's" and had no trouble finding another job. He spent one day "Working for my health," but the next day he was "Threshing at Davis's in the evening," and spent another six days there. His threshing job ended abruptly on September 19 when he "Got thrown from horse, and collar bone cracked." At this point, Pringle decided to go to his homestead. He spent four days travelling. He must have been greatly satisfied with what he saw, for his entry on September 23 revealed that he had done nothing but "Tramped around all day."[31]

Fred Pringle had achieved in one summer what he could not hope for in many years in the United States – entry to farm ownership. It was a meagre beginning, with unimproved land, no stock, no buildings, no implements, and no crop, but it was a start. At home, he had not expected to be able to climb the agricultural ladder. In Canada he did not have to.

The western Canadian version of the agricultural ladder was significantly different from its Ontario or American counterpart. The classic agricultural ladder, on which a man climbed from farm labourer to farm tenant and finally to farm owner, did not exist on the Canadian prairie frontier. Here, the ladder was significantly more flexible, and it could also be shorter. In western Canada, one could skip either the labour or the tenant rung. In fact, it was usually farm owners who became tenants, renting land after they had established themselves, as a low-risk form of expansion. Tenancy was often not an intermediary rung at all.

In his contribution to the "Canadian Frontiers of Settlement" series, R.W. Murchie sought to reveal the extent to which the agricultural ladder was operating. He scrutinized the agricultural progress of farmers in several farming districts and found little evidence that the classic agricultural ladder operated or was even necessary to achieve ownership. In the Swan River Valley, Manitoba, "A very large proportion of the settlers omitted the labour and tenancy stages and attained ownership directly." In Turtleford, Saskatchewan, 153 out of

178 farm owners had skipped the tenancy rung, and in Kindersley, Saskatchewan, only thirty-one out of 198 farmers had taken the route from farm labour to farm ownership via tenancy. In the Riding Mountain Fringe district of Manitoba, Murchie noted "the agricultural ladder has failed to function or . . . these farmers did not choose to use it."[32]

If tenancy was not a necessary stage toward farm ownership, then agricultural labour alone should have been able to provide the step up to ownership. The ladder on the Canadian prairies thus usually consisted of only two rungs. It might more accurately be called a stepping stone. The promises of farm work as prelude to farm ownership that had drawn men like Fred Pringle to the prairie West were repeated throughout the settlement period and well beyond. Promoters never carefully examined the accuracy of their assurances, and as a result there are few records available to document the extent of the phenomenon. Murchie's study has indicated, however, that the promises contained at least a kernel of truth.

Murchie found that although farm labour was not a necessary prelude to farm ownership, it was a route taken far more often than tenancy. In the Swan River Valley, forty-one out of 168 farm owners "went directly from positions as farm labourers to ownership," compared to only twenty-nine who had been tenants but not labourers. In Kindersley, 124 of the 198 farmers surveyed had worked as farm labourers and had not been tenants, compared to only eighteen who had been tenants but not farm labourers. But in all districts, numbers of farmers had achieved ownership without having worked as agricultural labourers at all. The proportions varied. In some districts, more than half of the farm owners had worked as farm labourers, as many as 63 per cent in Kindersley and 73 per cent in Turtleford. In Olds, Alberta, the proportion was just under half – 48 per cent. In other districts, fewer farmers had first worked for wages on the farms of others. In the Riding Mountain Fringe only 37 per cent had done so, and in the Swan River Valley the figure was even lower, at 24 per cent.[33]

The pattern of labour in prairie agriculture was thus not a simple distinction between waged work and farm ownership, with the latter negotiated through a series of stages beginning with the former. Newcomers might go directly from labour to ownership. They might homestead or buy land and work on the farms of others simultaneously and thus be farm labourers and farm owners concurrently.[34] Or they might even become farm owners before becoming labourers and use their farm wages to help establish themselves or to supplement their agricultural income.[35] In the long term, what is most significant about the agricultural ladder in western Canada is the ideological sway it held. Whatever the actual number of men who arrived at farm ownership through waged farm labour, there were enough who did so to convince others that it was not only possible but likely. In the short term, the significance of the western Canadian version of the agricultural ladder lay in the effects it had upon labour-capital relations.

Because a climb up the traditional ladder was not expected to take place overnight, it implied a lengthy and dedicated commitment to waged labour in agriculture and another long term as tenant. In the prairie West, the ability to skip the tenant stage, and simply to use agricultural labour as a stepping stone or even as an adjunct to farm ownership, severely undercut the willingness that men might have had in the older agricultural districts to look forward to long years of waged labour. In Murchie's study, farmers who took the steps from farm labourer to tenancy to ownership had taken "a substantially longer period" to arrive at their destination than did those who skipped a rung or two. In the Swan River Valley district, for instance, the average age at which present farm owners had begun working full time was fourteen years, yet the average age when they achieved ownership ranged from twenty-nine, for those who had skipped both the farm labour and tenancy rung, to thirty-two, for those who had skipped the tenancy rung and only spent time working as agricultural labourers, to thirty-three, for those who had skipped the farm labour rung and only been tenants, and up to thirty-six years old, for those who had climbed the classic ladder. In the Riding Mountain Fringe, it had taken farm owners an average of eighteen years to scale the agricultural ladder.[36]

Men who came to the prairie West to begin farming were impatient, reluctant to wait the years that a climb up the vaunted agricultural ladder might take. Percy Maxwell enjoyed his job as a hired hand, but twelve months was enough for him. "It is quite time I struck out for myself," he declared in 1904, "instead of spending the best years of my life working for somebody else."[37] This eagerness was underscored by the rush of settlement. Men had only to look around to see how quickly the land was being taken up. They could not afford to wait. On the other hand, if they were short of cash or lacked farming skills, they were well advised to spend some time acquiring both. This delay met the needs of both labour and capital. Farm workers needed cash and experience. Farmers needed cheap experienced labour. They reached a compromise. During the settlement period, farm workers and their employers entered into what amounted to an informal arrangement that can best be described as an apprenticeship program.

The system of agricultural apprenticeship was not a true apprenticeship, in which a master and a student enter a legal contract, the one to teach and the other to serve in the acquisition and performance of a craft. In prairie agriculture there was no formal contractual agreement explicitly defining rights and obligations, nor was there a fixed term or a recognized determination of skill acquisition. The only legal recognition was in the provincial Masters and Servants Acts that governed relations between hired hands and their employers. In Manitoba it also explicitly applied to apprentices.[38]

The arrangement did embody many classical features. Hired hands were generally younger than farmers but slightly older than farmers' sons, occupying an intermediate position appropriate to an apprenticeship. In 1911, for example, more than half of farm workers were under the age of twenty-five.[39] Most significant, however, were the opportunities apprenticeship offered. Men who engaged in farm labour expected to acquire the wherewithal to enter the rank of master themselves. Primary among their concerns was education in the craft of farming. The method of obtaining the necessary agricultural skills and know-how served both master and pupil, for it was a combination of instruction and practice.[40] The next important requisite to farming was capital, and this, too, reflected the usual arrangements between a master and pupil. Wages for farm work were notoriously low and during winter months often consisted only of room and board, a situation that was common in apprenticeships. But it was in the third requisite that the apprenticeship relationship was most apparent. The men who engaged in farm labour and the men who hired them fully expected that one day the relationship would end and the pupil would join the rank of the master. The social relations into which they entered were thus determined not only by their productive relations but by their anticipated equality of social and economic position. This social relationship, which is examined in greater detail in the following chapter, did much to mediate the inconsistencies in the arrangement and the variations in the conditions of labour.

Men who undertook farm labour as an apprenticeship shaped their work strategies to meet their own exigencies, striving to achieve a balance between what they were required to give to the system and what they needed from it. Farming skills were the most important because without them no amount of capital could make a man a successful farmer. On the other hand, men needed some capital to make a start, and those who entered the apprenticeship system found this the weakest part of the arrangement.

The glowing reports of immigration and settlement propaganda assured them that a short stint at agricultural labour would provide enough capital to make a good farming start. A comparison of a yearly farm wage with the costs of homesteading reveals that throughout the pioneering and settlement period this was at least statistically possible. In the 1880s, immigrants had been told that $300 was the minimum required to provide the barest necessities on a homestead and that $500 would provide a good start. A summer agricultural job that included room and board could pay from $25 to $40 a month, with an average of about \$30. At that rate, if a farmhand found extra work at harvest and room and board over the winter, and managed to save every penny, he could raise the requisite amount in two or three years.

But capital accumulation involves not just earning the money but holding on to it. The rosy translation of farm wages into homesteading costs failed to account for the expenses that even a frugal hired hand had to meet. On prairie farms it was usual to withhold wages until the end of the term of employment, with advances for tobacco, clothing, or an occasional night out. For farm workers intent on saving every dollar, this method of enforced savings could be useful. But even though they received room and board, farm workers faced unavoidable expenses that ate into their accumulated wages, lengthening the time, if not the amount, they needed to save.

Wages and homestead costs did not increase noticeably before the turn of the century. At that time, homesteading costs still represented about three years' farm wages, and the *Labour Gazette* blamed the shortage of farm labour on "the disposition of the farm labouring class . . . to become homesteaders." It found that such a course of action "was quite possible for a man after working for another for from three to five years."[41] But as the West began to fill up, costs to begin homesteading rose. By 1910, the recommended figure had tripled to \$1,500. Farm wages had risen, too, but at a much slower rate. At the annual farm wage rate of about \$225 a year,[42] homesteading costs represented six years of steady work. The war years appeared on the surface to present an ideal opportunity for farm workers to make the transition to ownership, as wages doubled and tripled from their pre-war rates. But, by this time, much of the good homesteading land had been taken, and costs of purchased land reflected wartime inflation. Still, men were encouraged to undertake farm labour as a vehicle to ownership.

It was not always necessary, however, for men to save the entire cost of setting up before they could begin farming. The homestead made this possible. In searching for an explanation of labour shortages in the North-West Territories, the *Labour Gazette* found that a "very potent factor . . . is the disposition of energetic men to homestead for themselves." They were not willing to continue as waged labourers when the opportunity to be farm owners was at hand, even if this did not remove them from the necessity of waged farm work. Rather, they found it "profitable to work out for at least part of the busy season in order to place themselves in a better position as settlers on their own account."[43] As Fred Pringle had discovered, ownership did not mean the end of waged labour.

There is sufficient evidence to suggest that immigration promises were not just good prairie salesmanship. It was never easy, but an experienced and frugal hired hand able to secure steady employment could, as the settlement literature promised, save enough to make a fair start on farming, as long as free land was still available. Homesteading could be carried on with a minimum cash outlay, if frugality and sacrifice were accepted as part of the bargain. Current research supports this observation, although there is disagreement about the speed with which farm labour could bankroll a homesteading venture. Lyle Dick argues that prior to the First World War, "a hired man usually earned enough to start farming on free grant land with a year's accumulated wages."[44] Dick's view is too optimistic. Calculations on paper support the notion that a year's wages could provide homesteading expenses, but this was only possible if men found steady work and had absolutely no expenses. More commonly, men needed the wages of at least two or three years, and usually more, to make a fair start, even when they planned to continue working out to underwrite their farms.

Murchie's examination of the agricultural ladder is of limited use in examining this question, since it does not distinguish the date at which men took up farming, even though it does show that their stint at waged farm labour was lengthy. A series of questionnaires completed by Saskatchewan pioneers dealing with farm establishment at least corroborates that the route to ownership through farm labour was both possible and easier while homestead lands were still available. Farmers were asked whether they had worked as agricultural labourers in Canada before becoming farmers themselves. Of the 178 who responded to this question and who had come to Saskatchewan between 1890 and 1914, forty said they had. But as homestead lands were taken, the number of settlers who travelled this route declined. The surveys document the change. Of the men who arrived in Saskatchewan between 1890 and 1899, 20 per cent worked first as agricultural labourers. Arrivals in the 1900-to-1904 period showed a much higher rate of 31 per cent. Thereafter the rate declined, with 21 per cent of arrivals in the 1905-to-1909 period and 17 per cent of the 1910-to-1914 arrivals working as farm labourers before becoming farmers.[45]

By the eve of the First World War, prospective homesteaders faced disappointment. In a report on unemployment in 1914, Saskatchewan Premier Walter Scott summed up the experience of immigrants who had been lured to the West:

Their vision, inspired by immigration literature of a small farmstead, a comfortable home and practical independence in a few years, were indeed quickly dispelled upon their arrival in the West. They then learned that good homesteads were some distance from centres of population, and above all, that it required considerable capital to take up homesteading properly.[46]

While a man certainly needed money to start, no amount of capital would ensure his farming success if he lacked the skills and expertise. Experience was regarded as the best teacher. Immigration pamphlets encouraged newcomers to learn Canadian farming methods first-hand by seeking work on Canadian farms. Letters home from recent immigrants were full of plans to work out in order to learn the ropes, and letters and diaries detail the process. "The main reason I took this job," recalled Willem de Gelder, a newcomer from Holland, "was to find out how to make this land as productive as possible."[47] Despite his agricultural background, de Gelder found that he needed to learn specific skills, too, such as the intricacies of the western harness and handling horses on the prairie sod. He was unused to his employer's method but determined to learn. "He works his horses in tandem," remarked de Gelder. "I've never handled horses in this fashion. . . . Well, I'll learn about it from close-up: it will be a little strange at first but I think I'll get used to it."[48]

Men eager to learn prairie farming methods sometimes faced financial sacrifices. Harry Self, immigrating from Liverpool in 1903, at first took work on the Canadian Pacific Railway line for several weeks but left it when he found the opportunity for agricultural labour. Self described his encounter at Pense, Saskatchewan: "I met a farmer at lunch time at the hotel who wanted a hand so, as my intention was to do that kind of work, I hired for the summer at $15.00 a month and board." Although he had been raised on a farm in England, he decided to spend three years gaining local experience. He took a cut in pay to do so. At the end of the summer, Self stayed on over the winter for room and board, and in the spring agreed to a full year's engagement at $200. But agricultural labour was a poor vehicle for capital accumulation. The next year found Self working for a farmer near Wascana, Saskatchewan, who suffered from ill-health and was unable to do much work. The advantage for Self was that he was put in charge of all the work on the 320-acre farm except driving the binder during harvest, the one job his boss was able to manage. This entailed hard work beginning every day at 4:45 a.m., but Self was satisfied with the arrangement. What he lost in wages he gained in experience, and by the following spring he was able to take out a homestead himself, right on his own schedule.[49]

Farm workers were very conscious of the benefits of first-hand experience, and they took advantage of the general shortage of labour and the wide variety of farm enterprises to increase their farming knowledge. Gaston Giscard was typical. He came from France with an agricultural background but knew that conditions on the Canadian prairies were very different. He regarded his farm work primarily as an opportunity to learn and took steps to ensure that he achieved his goal. "After apprenticing for a few weeks," he recalled, "I tell my boss that I intend to leave him. He's sorry that I'm going. He's surprised, too." Giscard's employer attempted to keep him by offering a higher wage, but Giscard was working to his own agenda: "I hasten to tell him that that's not my reason for leaving. I just insist on changing bosses to compare the different methods used by each farmer."[50]

Used as an apprenticeship, agricultural labour could enable a farm worker to make the transition to farm owner. It required men to employ strategies to take advantage of the particular conditions of settlement agriculture. It also required hard work, determination, flexibility, and a good deal of luck.

John Stokoe is a good example of a prairie newcomer who used the apprenticeship procedure to negotiate his way from labour to capital. Stokoe emigrated from England in the spring of 1903. He had gained some farming experience in England and planned to homestead in the North-West Territories, but first he wanted to learn western Canadian farming methods. While looking for work at Stonewall, Manitoba, he impressed a prospective employer with his enthusiasm. The farmer hired him on the spot and gave him a great deal of responsibility right from the start. "I am doing all the work just now," Stokoe proudly wrote in his first letter home. "My work is never supervised." He took the responsibility seriously, telling his family of his long day: "At night after the Boss & his wife have retired I take a lantern & go around the whole place to see that all is secure, & am up first in the morning, without being called, light fire & put on the kettle then go & feed stock, come back & replenish fire & sit down & read till someone comes down to make breakfast."[51]

Stokoe relished the opportunity to adapt his English farming skills to Canadian methods and was optimistic about his ability to learn. "At the rate I am going," he wrote confidently, "I will be able to manage this farm myself in a few weeks time, or any other farm." He ascribed his rapid schooling to the fact that he had been given "complete control of everything since I came." Stokoe also attributed his progress to his eagerness to learn quickly. "A thing is never explained to me," he wrote. "I just go by what I see." He impressed the boss's wife, who "reckons she never saw anyone pick up things so quickly," and the boss, who "said I had done everything to a T."[52]

As an entry into the agricultural community, Stokoe found farm work invaluable. By late summer, he had moved to a farm near Wood Bay and quickly made himself part of the community. "You see, while working on a farm," he explained, "a man can learn something of the country, courage, people & conditions of work." Even more important was the ready acceptance of an obviously ambitious man into the farming fraternity. He sought the acquaintance of established farmers and eagerly solicited their advice. "Social conditions over here have made it quite easy for me to become intimatley [*sic*] acquainted with big farmers," he said. He was welcomed, too, into the agricultural community, helping to form a local literary society, joining the church choir, and attending dances, picnics, and other social occasions. He was proud to be an accepted member of the district, identifying strongly with his new home and his new social relations. As he proudly wrote to his father: "We are a go-ahead people in Wood Bay."[53]

Stokoe's apprenticeship was obviously off to a good start, but it proceeded more slowly than he had anticipated. He remained at waged farm work for another three years, then, instead of homesteading further west, he purchased a farm. The delay and change of plans were not due to problems with his apprenticeship. Rather, it had worked so well that he had come to feel a part of the community, and he had been so successful at a wide range of farming ventures that he believed a purchased farm in a settled community was a sounder economic investment than a homestead, which would not begin to produce a reasonable return for at least three or four years. Financially, though, his years of agricultural labour were ill-rewarded. After three years of steady work and stringent saving, Stokoe had accumulated an average of only $120 per year. Part of this had come from his supplementary farming ventures such as stock-raising and sharecropping. He was able to raise the $500 downpayment for his farm, but he would be left with nothing to carry on farming operations. "Now, here's where I stick," he wrote to his father. "If I give him that money I won't have horses, seed, feed or anything else, I will be cleaned right out." Although he felt that his progress was cause for pride, Stokoe was reluctant to ask for help. "It was not my intention when I came out here to take any help from you if I could help it," he said, "but I didn't expect to run against these circumstances."[54]

For John Stokoe, the apprenticeship system worked. He had learned a great deal about all aspects of farming and had been welcomed into agricultural society. If he had been unable to enlist the financial help of his family, he could still have acquired a farm of his own through homesteading. The meagre wages he had been willing to accept were his part of the bargain. For the valuable experience he was gaining, and for the entry into the farming community, he felt well paid.

Stokoe is an example of the prairie farm hand in transition. In negotiating his role from farm labourer to farm owner, he occupied a number of different economic positions. His ability to employ strategies that enabled him to pursue his chosen course reflected the broad reality of prairie farming. Men who engaged in farm work were tenants and owners, farmers and sons, and waged agricultural labourers. There was fluidity among the positions, upward as well as downward, and the role of employee was not confined to men who were permanent paid labourers. In no other industry could men confidently expect to move up from employee to employer, still less from waged worker to independent owner. Throughout the pioneer period, in both the economic and the social realms, prairie farm workers occupied a unique and ambiguous place.

5

Class, Culture, and Community

". . . everyone was welcome – that was understood."[1]

After a peripatetic seven or eight years in South Africa and India, Charles Fisher was ready to settle down. His search for a promising future led him to Hazelcliffe, Saskatchewan, a new community that in 1907 boasted a "store and post office, the essentials, and one grain elevator."[2] Fisher was impressed by its potential. Here was a community in the making, and he was eager to be part of its growth.

His first steps were a two-mile walk to his new job at Albert Millham's farm, where he immediately sat down to a large meal. When Fisher admitted his surprise and embarrassment over the huge appetite he had developed, he was assured to hear that his was "the usual condition of new arrivals in the country." Fisher began to feel at home.

His plan for his new life was straightforward, similar to that of thousands of others who came before and after him to the prairie West: "I settled down to become a Canadian farmer-in-training as a 'hired man', as I intended later to get a free homestead." His apprenticeship in farming progressed well, and he soon found himself "more than pleased[;] as a novice I was learning." Fisher worked hard and was soon able to report that he had "gained knowledge of the many and varied needs and skills required of a farmer."

He also gained knowledge of the particular relationship between farmers and their employees. From the start, Fisher referred to his boss as "the man with whom I was to work," recognizing that labour-capital relations in prairie agriculture were different from those in other industries. Even though he rankled under some of the conditions of his labour, objecting to a workday that lasted "until eight or nine in the summer" and the "disability" of no minimum wage, problems he thought could be rectified by "labour [getting] organized," he took it for granted that the long hours, and steady work were necessary, and that even the custom of loaning out hired hands to other farmers was "a sensible practice accepting the inevitable fact of scarcity of labor." He noted matter-of-factly that a hired hand was "considered part of the farm. His employer's interests were those of himself."

Fisher was glad of the opportunity to learn Canadian farming methods in an atmosphere of relative partnership with his employer, but it was in the area of social relationships that he felt most at home. His introduction to the local community began through the world of work. He quickly "made acquaintances with men interested [in and] well read on subjects . . . of wheat from infancy to maturity." He shared their concerns about frost and harvesting, and became a staunch advocate of hail insurance after witnessing a hailstorm ravage an "entire growing crop of wheat, smashed to the ground in a matter of seconds, one year's work lost entirely." He worried and debated about marketing and grain companies, about railway monopolies and roads, concluding that the "grievances of the farmers were very real."

In the agrarian community where Fisher had taken up residence, the concerns and interests of farmers dominated the economic sphere and directed community and social development. Fisher was struck by the spirit of co-operation evident at all times, but especially during the harvest season. Threshing was "a gathering of all and sundry to cope with the work. Neighbours helping neighbours with their hired men." The social dimension was equally important, for the gathering "broke the spell of isolation, suffered the greater part of the year."

Fisher found himself a welcome addition to the social life of the community and was invited to all local events, from picnics to dances to card parties to religious services. He noted that the "paucity of population, the distances apart," and the steady hard work helped create a spirit of co-operation and comradeship among all members of the rural community. Religious services were followed by "get togethers," work bees were followed by dances, and the summer picnic was the high point of community co-operation: "all the people of the area there prepared to enjoy themselves for the day with games and races, etc. The idea of providing food was a community effort, all brought food baskets with them which was put into the common pool." Fisher entered into the entertainment with relish, and when it was over, "his joys and sorrows finished for the day, tired and happy, [he headed] back to the farm for the usual chores."

Fisher's welcome into the local community was not determined solely by the shortage of population. When he dropped in on neighbours he was welcomed "heartily and not [as] a formality by any means. Truly genuine." As an ambitious hired hand on the way to becoming a farmer, Fisher was assured a place as a productive and respected member of the dominant agrarian economy. But he also had the advantage of belonging to the dominant ethnic group.

Fisher's first taste of ethnic relations in the prairie West came when he learned that "the 'welcome mat' was not put out for Englishmen." He was surprised and angry that their lack of prairie farming expertise should saddle all newcomers from Britain with the label "Green Englishmen." But he soon discovered that animosity was even stronger toward immigrants from central and eastern Europe, who he reported were in the "non-welcome category," toward "coloured people [who were] relegated to the damned," and toward "our original Canadian stock [who have] had little love bestowed upon them."

Anglo-Celtic culture dominated the prairie West, and the supposed ability of immigrants to assimilate to that culture did much to determine the degree of their social acceptance. European immigrants, particularly those from southern and central Europe, laboured under heavy burdens. Fisher echoed a common resentment when he thought that such immigrants might have "been assisted by the government to become established with flour and other needs." He expressed a common fear when he discovered that immigrants soon learned that "'Me Liberal'" was a "magic password" to facilitate the homestead filing process.

But faced with the economic realities of a sparsely populated and struggling community, Fisher developed an appreciation for the contribution that the often unwelcome immigrants made and for the trials they faced. He sympathized with the plight of those new arrivals who had no resources to tide them over the first winter and who existed in "a state of 'financial Strengency,' [sic] adding to their troubles." He blamed the "slimy acts of some politicians" for magnifying "racial distinction[s] . . . out of all proportion." And he understood the ways ethnic differences affected both sides of the employer-employee relationship: "with few exceptions [the immigrant] was alike in trying to adjust himself to new customs and conditions imposed upon him[,] plus his employer might justifiably be exasparated [sic] upon occassion [sic] even to admitting that he knew it was raw material with which he was dealing."[3]

Fisher was well aware of the complexity of social relationships in the developing prairie West. Lines of class and ethnicity were seldom clearly drawn or rigid. A neighbour might be an employer or an employee, a member of the dominant ethnic group or a newcomer from an unwelcome minority. But the economic and social needs of the developing community determined that all would exert some influence in shaping the emerging agrarian community. Hired hands like Charles Fisher had their own niche.

Farm Workers and Prairie Culture

Studies of working-class culture demonstrate how a strong working-class community can nurture proletarian behaviour and develop an institutional framework to provide the support for collective action.[4] Prairie hired hands enjoyed no such community support. Villages of farm workers were not successfully established in the prairie West, despite various suggestions and even attempts by colonization and land settlement companies. Simple logistics were part of the reason – hired hands for the most part lived on the farms of their employers, precluding any community network of support. More important were the ambitions of farm workers to move quickly beyond the stage of waged labour. As apprentice farmers, they sought to establish links with the class to which they aspired rather than with the class to which they belonged, and they subscribed to an ideology and culture that represented their aspirations rather than their realities.

An examination of class and culture among agricultural labourers poses difficult questions. Hired hands found themselves in a situation quite unlike those of workers in other industries, and their responses were likewise geared to their own circumstances. It is not possible to arrive at a simple characterization of their lives or of the social and economic milieu in which they operated. Moreover, historians have been far from agreement in their analysis of these issues. The question of class is especially problematic. Historians continue to debate the economic nature of agriculture, and few are willing to see farmers as capitalists. Indeed, for an industry characterized by such a high degree of labouring ownership, it is difficult to draw clear lines. The question often comes down to an interpretive decision about the nature of independent commodity production.[5]

When hired hands enter the scene, the issue becomes even more problematic. Few historians have addressed the question of the class position of farm workers. Those who do are not in agreement. In my own work, I have argued that by the 1920s farm workers were a proletariat. Gerald Friesen maintains that agricultural labourers were members of the working class, but they did not exist as a working class for any length of time. Lyle Dick draws upon Michael Katz's industrial model and says that "farm labourers did not conform to the usual definitions of class" because they did not engage in open class conflict and because of social interaction between themselves and their employers. Paul Voisey argues that class was seldom a factor in conflicts in the West. W.J.C. Cherwinski avoids characterization. John Herd Thompson and Allen Seager have found proletarian behaviour among sugar-beet workers, but that group and others who work on large labour-intensive agricultural enterprises are beyond the scope of this study.[6] Here, the focus is on the lives and class position of workers who did not establish an easily identifiable connection with other workers.

Hired hands became enmeshed in a culture and an ideology that were in opposition to their position, and ultimately, to their needs. "Ideology" here means "the set of ideas which arise from a given set of material interests," a definition derived from Raymond Williams[7] but further refined in Chapter 8. "Culture" is used here in the Thompsonian sense, to include such broad categories as "traditions, value-systems, ideas and institutional forms."[8] The term "dominant culture" is more restricted and does not imply hegemony. Culture is adaptable and in flux. It is an arena of continuing class struggle rather than a process by which one culture imposes its own ends and its own vision on another, thus reducing the other culture to a kind of corporate culture that is essentially self-defensive, although it may seek to improve its position within the social order that has been determined beyond its control. Proletarian culture is just as purposive as bourgeois culture in its endeavour to perform a transformative role in society, even though the former is less successful than the latter. In this study, the term "dominant" refers to the culture that has achieved pre-eminence over another.

The dominant culture in the prairie West was that of capital rather than labour, and it derived, in the rural setting at least, from the economic importance accorded to agriculture. There was "no one universal prairie culture," as John Herd Thompson and Ian MacPherson have pointed out, precluded by differences in "ethnicity, religion, time of settlement, varied agricultural practices, and the differing economic opportunities offered by local environments."[9] Nonetheless, the values and institutions that emerged from and directed rural society were those that reflected the economic base and social attitudes of the dominant agrarian group, the farmers. And as prospective farmers themselves, hired hands occupied a special niche. Within the social structure of the farming community their position was one of apparent equality.[10]

The location of hired hands within the dominant culture obviated their need to develop ties with the working class, even in a cultural sense, and indeed militated against it since their very status was contingent on their participation in the activities and institutions of the dominant culture. The wage labour aspects of their lives were downplayed. Farm hands seldom felt the need to establish links with the working-class world in the present, nor did they anticipate the usefulness of such links in the future. Hired hand Fred Wright demonstrated how he sided with capital, rendering a harsh judgement on a group of harvesters who had been holding out for higher wages. "They had no idea of working," he observed. "They were called Industrial Workers of the World and we also called them 'I Won't Work' and they didn't.

They were a hard bunch to deal with."[11] It was a circular process: once situated within the dominant class, hired hands were absorbed into its culture and in turn absorbed its values and outlook.

The cultural structures that gave shape and continuity to the farm workers' lives were thus determined by the farming community and reflected the interests and direction of the leaders within that community, the farmers. While working as hired hands, George Shepherd and his father found themselves drawn into "the spirit of the frontier . . . as we became more conscious of farmers' problems and the farmers' point of view."[12] The associational network to which farm workers belonged was centred in the rural community. The names of hired hands can be found on the membership lists of local agricultural societies, whose direction was always firmly under the control of the farmers.[13] Other types of organizations were as diverse as debating societies and hunting clubs.[14] Such organizations were not antipathetic to agricultural labourers, but neither did they provide an environment in which farm workers could identify a shared outlook or develop their own cohesion as waged labourers.

Indeed, the self-conscious if superficial democracy of pioneering society decreed a social structure that was not stratified on class lines.[15] Historians have grappled with the question of class structure in the prairie West, acknowledging social hierarchies based at least in part on economic status but disagreeing on the importance of such stratification.[16] Distinctions were often based on criteria besides occupation or financial standing, such as ethnicity, religion, and politics. Personal characteristics such as industry and perseverance were often even more important, especially when homesteaders began farming in an undercapitalized state and were heavily dependent on wage labour themselves.[17]

Consequently, hired hands became firmly entrenched within the dominant social structure of the farming community. The ease of their social acceptance is reflected in pioneers' reminiscences. Farmers and other members of the agricultural community have asserted that the hired hand was treated as one of the family and welcomed into the farm society, participating fully in sports events, entertainments, and recreational and religious activities.[18] Hired hand Frank Timmath of Lavoy, Alberta, sums up the contradiction. Timmath acknowledged the contract between labour and capital as well as the shared animosity to the farmers' common foe:

> Now John, I ask this much of thee.
> I've tried to serve you faithfully.
> If I should die here, from home afar,
> Don't ship my remains on the C.N.R.[19]

Acceptance into the farming community meant that hired hands were constrained from establishing cultural ties with other members of the working class. Among themselves, they fashioned a loose cultural connection rooted in a common ambiguity in their relationship with capital and in a shared response to the conditions of their labour. But this did not lead them to abandon their place in the dominant culture of the agrarian West.

Farm Workers and the Construction of Gender

A measure of farm workers' acceptance into the dominant culture, and an example of the social side of the apprenticeship system at work, is seen in a closer examination of one of the most important cultural institutions in prairie agriculture – marriage. Its economic and social significance illustrates an intersection of themes affecting the position of hired hands. Recent studies on the construction of gender have plotted the complex relationship between gender and class.[20] In the case of prairie farm workers the correlation is particularly clear. Masculinity was an important component in the package that mediated labour-capital relations. The physical traits of manliness were affirmations of self-worth in a society that valued hard work and offered little financial reward. But masculinity in prairie society and culture had another dimension. Most hired hands were bachelors.

The agricultural community had been designed and was defined, both economically and socially, as family-oriented, based on small-scale units of production – family farms. The economy provided the infrastructure for the society.[21] The advantages were obvious. The economic contribution of a family was proportionally much greater than their mere numbers, since the costs of their labour and provisions were hidden in their production. Politically and socially, individual farm ownership meant conservative values, while the predominance of families ensured the entrenchment of institutions and fostered social stability. Westerners embraced the vision of the family farm as the "ideal social unit."[22] Yet there was a well-defined niche for bachelors.

The term "bachelor" had a particular connotation in the prairie West. Its basic designation, of course, was an unmarried man. But it also referred to married men who were temporarily without their wives. Elizabeth Mitchell, a careful observer of western Canada before the First World War, gave the following definition: " 'Bachelor' has the technical meaning of a man living by himself or with other men, with no woman in the house. A widower or grass-widower 'batches', an unmarried man with a sister or housekeeper does not."[23] Married men often found themselves in this condition. When Wilfred Eggleston's family decided to begin farming in 1909, his mother and the three children stayed in Nanton, Alberta, for school while his father took up a homestead. Young Wilfred observed that the family looked forward to their reunion, but "In the meantime my father would have to rough it and 'batch it' on his own."[24] The term "to batch" entered the prairie lexicon with ease, indicating the social acceptability of men without women. In some ways this is a little surprising, since both the social and economic progress of men on their own lagged behind those of men who had their wives and families at hand. Bachelors were perceived to live in squalor and loneliness. Their homes were "wretched establishments," even by rough pioneering standards.[25] Bachelor shacks were makeshift, thrown up quickly to provide little more than the barest shelter to fulfil homesteading requirements. They were unpainted, cramped, and had no amenities.[26] Prairie bachelors bemoaned their condition. Young Edward ffolkes described the hardships of single life. "The bachelor lives on pork and bannocks, as a rule;" he wrote to his mother, "never sweeps his house out, or very seldom; generally hoes the floor once a month."[27]

The economic progress of bachelor farmers was expected to be much slower than that of their married counterparts. Batching on the prairies was "the most expensive way" to keep house, explained ffolkes, "because [the bachelor] has no time to make bread often, or even butter, in summer, or puddings, or soups with vegetables, which saves the meat – and meat is expensive."[28] The *Nor'-West Farmer* was only one of many voices to counsel the acquisition of a wife as a necessary condition to economic success. In reply to a bachelor's plan to begin raising livestock, it warned: "one man on a farm can hardly make a success in mixed farming. Better look out for a female partner, if you can arrange to that effect, and plan at the same time for heifers to come in at between two and three years."[29]

Isolation increased the hardship. In an item on the mental illness of a local homesteader, the *Hanna Herald* observed the general self-neglect that resulted from a solitary existence on an isolated farm. "Many a lonely homesteader puts in his hard day's toil and retires at night on a meal quickly made by his own hands and scant." The problem could be apathy or exhaustion: "after working in the field he has not the inclination to go to the trouble of preparing a better [meal]."[30] Bachelor homesteader Ebe Koeppen from Germany had first-hand experience of the toll of isolated living. He recorded in his diary that he had reached a "very sad point." Life without a wife was "slow suicide. Slow spiritual death." He confined his observations to his diary, explaining "I do not write home about these things" because the "staggering drearyness [*sic*] of such existence is too difficult to make understandable."[31]

Bachelors lamented their fate, although many tried to make light of it. A popular prairie song, "The Alberta Homesteader," described "Dan Gold, an old bach'lor" who was "keeping old batch on an elegant plan." Prairie bachelors could empathize with his miserable conditions:

> My clothes are all ragged, my language is rough,
> My bread is case-hardened and solid and tough
> My dishes are scattered all over the room
> My floor gets afraid at the sight of a broom.[32]

But these very tangible impediments to economic and social progress did not cause bachelors to be shunned. Quite the contrary. Bachelors were viewed with tolerance, concern, and benevolence, and often with bemused affection. Westerners, particularly women, felt a certain responsibility toward them. Part of women's duty in the West was "being kind to poor bachelors round about who need kindness badly."[33] Bachelors were more to be pitied than scorned, and the sorrier their existence, the greater the solicitude.

The men developed strategies for dealing with their unhappy position, and the local community willingly met their needs. Recalling their first winter on their homesteads, Koeppen and his homesteading partner Hans Boske admitted "we damn near lived on rolled oats and corn syrup." But they soon became acquainted with the nearest family, who were "lovely people. The wife would bake bread for us once a week and we would work it off."[34] This kind of response was common, and clearly demonstrated that there was a special niche reserved for men without women. "I used to feel so sorry for those boys," recalled one pioneer. "They were so pitiful. My mother, she worried over them and she babied them and they came to her with all their troubles."[35]

Aside from compassion, there were other more practical reasons for welcoming bachelors into the family-oriented community. Despite the delays in their progress, they did make tangible economic and social contributions. A simple shortage of population meant that all settlers were eagerly included in all economic and social functions. Welcomed at barn-raising bees and on harvest crews, at dances and at ball games, prairie bachelors pulled their weight in the developing agrarian community. In 1885 the *Qu'Appelle Vidette* reported a dance hosted by "the bachelors of the northern part of this municipality" for a crowd of about forty. An "excellent repast" began at six, and dancing went on "until daylight warned the delighted but wearied party that it was time to homeward wend." The bachelors had done themselves proud and earned the admiration of the community. The *Vidette* summed it up with the observation that "it is doubtful if anything better could be turned out anywhere in the Northwest, outside of Winnipeg."[36]

Bachelors could also be easily forgiven for their single state, since the prairie population was still very small and badly skewed in its male-to-female ratio.[37] Pioneer David Maginnes of Balwinton recalled a dance with only three women to thirty men: "We danced until the women got tired."[38] C.A. Dawson and Eva Younge's survey of ten farming areas shows that the newly settled areas of Turtleford, Saskatchewan, and Peace River, Alberta, had a rural male:female ratio as high as 202:100 during the first decade of settlement. In older, stable settlements such as Red River, Manitoba, the ratio was 121:100 even at the beginning of the century.[39]

But the major reason bachelors found such a ready acceptance in the family-oriented community was because their condition, like many other pioneer hardships, was perceived as temporary. Felix Troughton, who had been a young bachelor on the prairies during the early pioneer period, entitled his memoir *A Bachelor's Paradise*, but his description tells another story. He drew upon an unnamed western farm journal to paint a gloomy picture of the prairie bachelor's existence:

> The young man rises in the morning, leaves unmade the bed that has not perhaps been made for weeks, he then feeds his oxen, and in the unswept and dusty house prepares a hasty and ill-cooked breakfast, which is eaten from off unwashed dishes. The bread is generally sour, hard or dry, the butter salty and rancid, the coffee worthless, the meat burned on one side and raw on the other. The breakfast table is left covered by dirty dishes and slops, and a million flies gather to feed in undisturbed peacefulness. The unrefreshed bachelor goes to his work lonely, miserable and dyspeptic. At noon he unyokes his oxen and turns them loose to feed, then goes to his shack and makes a hot fire in order to get his dinner. Heated by hard work under a blazing sun, a good wash, a cool room, and a well-cooked meal is what he requires, and what he cannot have. At supper, it is the same thing over again. His underclothes, seldom washed, become clogged with perspiration, and his bedclothes are in the same unhealthy condition.[40]

It was not only his person that suffered, but his farm as well:

> When he is away at work, the hawks soar around the forsaken house and catch the chickens in the yard; the pigs get into the garden, if he has one, and the calves get out of the enclosure and suck the cows. Sometimes the house is burned down from a spark that may drop from the neglected stove. When the bachelor has to go to town for supplies, mail, etc., cattle get into his grain fields, or pull down his stacks, and there is no one to let the dog loose, so the marauders riot at will undisturbed.[41]

Where was Troughton's paradise? Like so many other prairie promises, it was in the future. It was free land, and the promise of economic independence, for the "average plucky, physically fit, and red-blooded young man." But to achieve this independence, the "Bachelor Farmer," of whom there was "no more amiable, industrious . . . and persevering" person, must first of all bring his bachelorhood to a close:

> What the bachelor requires in his home is a broad-shouldered, stirring wife, who will keep the house in order, as well as the husband who owns it, and who will see that clothing and bedding are made clean, and are kept so; who will serve a well-cooked meal with fresh, sweet bread of her own making; who will see that groceries are good, and that proper value has been received for money expended; who will wash and mend her husband's clothing, and will remove the shingle nails that have been used as substitutes for buttons; one who will look after the hens' nests, see that the dairy is kept in order, and who will place the Bible on the table when the day's work is done.[42]

With such a wife, the new husband's "bearing will be erect, his eyes clear, and his purse full, his garden will have flowers, and his shirt will have buttons."[43] Even more restrained observers such as Mitchell insisted that marriage "means leaving a ghastly loneliness for companionship and help, and squalor for decent comfort."[44] A hardworking, supportive wife could spell the difference between success and failure. "A man who would be a failure alone," declared prairie observer Marjorie Harrison, "will pull through if he has a good wife."[45]

Bachelorhood may have been regarded as a disadvantage, but not a permanent one. Married men who were batching were doing so only until they could send for their wives and families. Single men were expected to be bachelors only until they could provide for a family, or until the shortage of women in the West could be redressed. There were few "confirmed" bachelors on the pioneer prairies; most unmarried men were "eligible" bachelors. And given the rate at which "every girl is pounced on directly she puts her face inside the settlement,"[46] it seems that the eligible bachelors were eager to end this temporary condition. "Us bachelors were all desperate," recalled Ebe Koeppen, "you can use the word 'desperate' quite properly here – we were desperate to get a wife one way or the other."[47]

Just as bachelorhood was temporary, so the family-oriented economy and society of the West had not yet been realized. The reality of the pioneer years was that the population was overwhelmingly male. The idea of settlement was so strongly linked to masculine endeavour that the term "bachelor homesteader" came into widespread use. It was a positive designation, embodying many of the attributes needed on a frontier. Such men elicited admiration. Kathleen Strange, writing from the perspective of a "modern" pioneer, described the common view: "The first bachelor homesteaders had come out, most of them, with nothing more substantial than courage and optimism with which to battle against the harsh elements of a new country and a new life."[48] In the crucible of this new country and new life an identity of bachelorhood was forged, an amalgam of the idealized attributes of bachelor homesteaders. It provided an identity for the men who made up the group, and it was incorporated into a broader prairie pioneering ideology. This fusion enabled bachelorhood to receive a very positive assessment and gave the identity additional legitimacy and strength.

Bachelorhood was male, which meant that attributes ascribed to it would be considered masculine. At the same time, it would mean that traditional qualities of manliness would be incorporated. But gender identities are fraught with contradictions, and manliness in one context is not necessarily manliness in another. So, although bachelorhood was a masculine identity, it was a particular type of masculinity, one that served the needs of men who were young, generally of limited means, and engaged in small-scale pioneer agriculture on a relatively womanless frontier. Each of these elements was incorporated into the identity of the pioneer bachelor, which thus resonated with the ideals of the dominant culture.

Much of the definition had to do with physical prowess. Agriculture was a physically demanding occupation – it required strength, stamina, and physical dexterity. With a long tradition as affirmations of masculine identity and validations of male self-worth, these characteristics were readily accepted. Farmers measured themselves and each other by "the supreme manly qualities" of a man who "could keep up to two binders."[49] The needs of agriculture were thus well served by incorporating these characteristics into the identity of bachelorhood.

On the prairies, small-scale agriculture was being practised, with individual farmers working individual farms, making decisions and carrying out an endless round of tasks on their own. As a result, individualism was an important component of the identity. But the extreme of individualism – competitiveness, which is a frequent component of masculine identity – was less important in the bachelor identity, since it did not serve the needs of an agricultural community struggling to establish itself. Co-operation rather than competition was necessary for survival in the early years.

Pioneering called for a spirit of adventure, for courage and resourcefulness, for the desire to leave behind an old life and to embark on a new one. This life required a will to succeed, adaptability, and a willingness to subordinate present gratifications for future possibilities. Taking pride in craftsmanship and deriving satisfaction from a job well done, commonly expressed characteristics of masculine identity, took on added importance to pioneers who often saw little financial reward for their hard work.

Homesteading, too, helped shape the bachelor identity. It required thrift. The low filing fee meant that men with limited resources could begin farming, but to do so they had to be able to husband their financial resources, demonstrate ingenuity and resourcefulness in enlarging them, and be willing to endure hardships in the present in the expectation of comfort in the future. There had to be an acceptance, too, of an aggressive democracy that decreed that to be cash-poor was not dishonourable. The *Nor'-West Farmer* summed up the attitude toward farming on the frontier, quoting "one of the social teachers of the century," John Ruskin: "'There is no degradation in the hardest manual or the humblest servile, labor, when it is honest'."[50]

Some parts of the identity were derived in contradistinction to the identity of other groups. Bachelor manhood was measured in its dissimilarity from womanhood, and also in its dissimilarity from childhood, but these measurements were not derived in an antagonistic way. Women and children both served an important and highly visible economic role in prairie agriculture. Except in unusual circumstances their roles were complementary to those of men, so they seldom posed a threat to adult male occupations.[51] Although bachelorhood was the antithesis of femininity and immaturity, the maleness and the maturity of bachelors were neither defensive nor aggressive traits.[52] Rather than concentrating on their physical superiority, bachelors defined their contrast to women in their lack of such social niceties as an ability to take care of themselves, to cook, to sew, to keep clean, to converse politely, and to demonstrate other features of a gentle civilization. Their contrast to children was defined by steadiness both in physical work and in life commitments, maturity of judgement, and wisdom from experience. All were necessary qualities in agriculture.

By and large, bachelor homesteaders found their identity within the dominant ideology. They were expected to live a wholesome, frugal life in anticipation of the time they would leave the bachelor state to become full-fledged members of the family-oriented agrarian community. Likewise, for prairie farm workers during the pioneer period, their actual economic status mattered less in the social network than did their preparation for full economic membership in the agrarian community.

Ethnicity Enters the Equation

The agrarian community in the developing West had many variations. Government policies that fostered block settlements as well as newcomers' desires to find homes near others of their own ethnic or national backgrounds gave rise to group settlements and ethnic enclaves. As communities interacted, tensions could surface. In an age when ethnic and national differences were defined as immutable characteristics, the desire to populate the West with agricultural settlers often came up against a resistance to peopling it with those who did not fit into the cultural pattern of English Canada.

The close link between agricultural labour and land settlement exacerbated the tension. The ideal farm worker was one who would become an ideal yeoman farmer. He must be more than industrious and ambitious. He must also be willing and able to enter the mainstream culture. Tolerance toward farm workers who did not fit this mould underwent considerable variation over time. In the early years of settlement, when western Canada was plagued with the loss of more population than it could attract, examples of ethnic tension and conflict were rare, at least in agriculture. Most of the non-English settlers were able to find and provide work within their own communities. More significantly, though, there was such a need for farm help that newcomers of any origin were able to find work. This is not to say that farmers were eager to accept farm workers from outside their own ethnic or cultural group, but in a period of sparse settlement they had little choice. In 1887, J.S. Armitage, the agent for the Manitoba Department of Agriculture and Immigration, compiled a file of letters outlining his difficulties in providing labour to local farms. He resorted to urging farmers to hire men who might not entirely fill their requirements. "The only men here at present are Icelanders," he wrote to a Carman farmer, "they cannot speak English but are good workers. If you would like one of these please let me know and I will endeavor to get a good likely looking one for you."[53]

Faced with the realities of a scant labour market, farmers sought men who could do the job, looking above all for competent, and preferably experienced, workers. When Ukrainians established a settlement in the interlake region of Manitoba, they found farm work only with the nearby Mennonites. But as their reputation as skilled harvesters spread, they were soon able to find work among English-Canadian farmers.[54] Later Ukrainian settlers benefited from this experience.[55]

The overriding concern to establish agriculture on a solid economic foundation found its champion in Clifford Sifton, Minister of the Interior. His accession to the post in 1896 and his responsibility for immigration signalled the beginning of an aggressive drive to entice an agricultural population to the West. His famous dictum regarding the high value he placed on "stalwart men in sheepskin coats" was soon translated into a burgeoning and polyglot farm population. To Sifton, ethnicity was not an overriding concern. Nor, it appears, was it a conclusive factor in the labour calculations of farmers who, above all, were anxious to produce a crop and bring it in. When they looked for hired men, skill and experience were ultimately more important than ethnicity.

But ethnicity did matter. As the late nineteenth-century trickle of settlers turned into a flood, the growing numbers of highly visible central and eastern Europeans met with growing disapproval. In 1905, the ministry and its immigration campaign passed to Frank Oliver, a firm supporter of a more selective policy. At first, despite his own disapproval, Oliver continued the policy Sifton established.

Still, practical farmers of the West, as much as they may have deplored the cultural mix of new settlers, found themselves hamstrung by a labour shortage. Prejudices could easily be suspended if the demand for farm workers was strong enough, as in 1909 when the *Manitoba Free Press* reported that the farmers, "face to face with a serious labour problem," were accepting the "hitherto despised Doukhobor and Galician [Ukrainian] . . . with open arms these days, and the only regret is that there are not more of them."[56]

The solution, in Oliver's eyes and in the eyes of many westerners, was not to restrict the amount of immigration but to limit it to British immigrants or to those who were believed able to assimilate more readily into English-Canadian society. Oliver could do little but pursue his predecessor's publicity campaign as long as immigration policy remained inchoate. But immigration legislation enacted in 1910 pointed the way to a more restrictive policy.[57] Soon, informal directives warned immigration officers: "There are certain nationalities which are not suited to agriculture."[58] Immigration agents and inspectors at boundary points were given lists of acceptable farm workers: "the nationalities desired by us . . . include Finns, but not Austrians, Poles and Italians, whose immigration was not desired."[59] Germans and Scandinavians were welcomed, but not Spaniards. Special emphasis was placed on keeping out Asians of any nationality, as well as "colored labour."[60]

Since permanent settlement was the uppermost aim of immigration policy, restrictions were applied even to temporary workers who might become permanent residents. Agents were instructed to be discreet but firm and to use their "very best judgement" in denying entry to such workers.[61] When black farmers sought refuge in Canada in the face of racial discrimination in Oklahoma, public policy supported popular opinion in taking steps to prevent their entry.[62] When a group of "fifty coloured labourers" sought harvest work in the West, they were forestalled: "I would not recommend you and your friends to come to Canada for the harvest, nor do I think this country would be suitable to you as a field for settlement."[63]

By the end of the first decade of the twentieth century, the rapid growth and ethnic diversification of the prairie population lent credence to those who feared that Anglo-Canadian culture and institutions would be challenged by the visibly different newcomers. Fears were voiced about the nature of Canadian social development. Infant prairie communities were torn between the desire for a growing population and a need for labour, and the fear that community development might take an undesired direction.

Immigration agents responded by attempting to restrict "undesirable" groups. Farmers responded by restricting their offers of employment. In a growing labour market, farmers could begin to exercise their preference for Canadian or American help. But in farming communities, the question was never simply one of ethnic desirability. Farmers above all sought competent help. They were blinkered in their assessment of the capabilities of farm hands from eastern and southern Europe – "farmers will not touch Italians with very few exceptions"[64] – but were practical enough to recognize the necessity of hiring labour where they could find it. Indeed, even ethnically desirable British immigrants were sometimes faced with the warning that "No English Need Apply." Farmers were suspicious of "green Englishmen," believing they were infused with the same reputed laziness and farming incompetence as remittance men or were unemployed urban dwellers, unable to make the adjustment to the hard work and long hours on prairie farms.

When war broke out, intolerance became entwined with nationalism. Tensions rose. A number of "undesirable" immigrant groups found themselves subject to hostility as enemy aliens. And the link between labour radicalism and some ethnic groups that had developed before the war increased the Canadian fear and antagonism. Intolerance reached new heights in the later years of the war, despite a labour shortage that reached crisis proportions. Blacks, even as harvesters, were systematically excluded from entry, and the suggestion that Asian labour be brought in was met with demands for the strictest controls: "it should be under military supervision, should remain under military supervision, and finally be deported under military supervision."[65] Farmers were caught up in this growing antagonism, yet they were still primarily concerned with the practical aspects of farming. These conflicting concerns came together in the view that enemy aliens must accept farm work at low wages to prove their loyalty.[66]

The post-war world ushered in a new stage of relations between immigrants and residents. Antagonisms persisted, fuelled by the growing proportions of central and eastern Europeans in the West and by the realization of a full, indeed, glutted, labour market. It is impossible to determine if farmers were reluctant to hire central Europeans out of simple prejudice or because they did not believe that such workers had the necessary farming skills, but in the full labour market of the 1920s they were able to exercise their choice. Of the 200 applications for farm help the Lethbridge office of the Alberta Employment

Service received in April, 1927, "Not one single order has been sent in calling for the [central European] type of man."[67]

By the 1920s, farmers could afford to be selective in their hired help. Records of the colonization department of the Canadian Pacific Railway amply illustrate this. The CPR, like many other land sales agencies, was actively involved in finding farm jobs for potential land purchasers. In its capacity as a labour placement agency, the CPR took applications from farmers and tried to fill their orders from among immigrants who had come to Canada under its auspices. Applications were often very specific, detailing the type of farm help wanted according to experience, age, nationality, religion, marital status, and size of family. Farmers were also required to specify length of employment, type of accommodation, and wages the farm worker could expect. From among the applications collected in southern Alberta between 1927 and 1930, a sampling of 574 has been examined.[68] They reveal not only the type of farm worker sought but variations in wages and working and living conditions.

Those of British origin were most in demand. Nationality or language was usually specified, with a 50 per cent preference for British, or at least English-speaking, workers. Scandinavians were requested in 32 per cent of the applications and Germans in 10 per cent, indicating a clear preference for workers from northern Europe. Less than 4 per cent requested farm help from eastern European countries, although another 4 per cent were willing to accept workers of any nationality.

While these preferences are hardly surprising given the ethnic make-up of southern Alberta's population in the late 1920s, it is noteworthy that the workers who met these specifications could expect steadier work and higher wages. Permanent or full-time jobs were offered in only 16 per cent of the requests for eastern Europeans but in 36 per cent of those for Scandinavians. The work, too, was less onerous for British or northern European farm hands. When a farmer at Benton asked for workers, he wanted men "for removing rock from land so I think that Hungarians might be best."[69] In more than half the applications for eastern Europeans, the work offered was clearing brush and hauling rocks. Such jobs were seldom offered to northern European or British workers, who were most often expected to undertake "general farm work." Wages were similarly adjusted. The average monthly wage offered to British or English-speaking farm hands was $35 and for Scandinavians it was $30, but for eastern Europeans it was only $25.

The rosy picture immigration agents painted clearly had problematic dimensions for immigrants of "undesirable" ethnic and national affiliation. The repercussions for western development have been well documented.[70] Farm workers frequently fared better among their own countrymen, according to immigration agents, who found that "foreign residents at country points will take a much greater interest in their compatriots than will English speaking residents in the British worker."[71]

Ethnicity played an important role in the relationship between farmers and their employees and among farm workers themselves. These relationships, where they help to elucidate labour-capital relations, are examined in subsequent chapters. As long as men sought farm work within their own communities, ethnicity was not a factor in their relationships with their employers. But when they sought work in other communities, they encountered a mixed reception from employers and fellow workers alike.

Questions of class, culture, and community pose special problems in the developing West. Farmers dominated the economic scene, and their values did much to determine social relationships. But farmers were a far from monolithic group. They were cash-poor homesteaders and prosperous agriculturalists, bachelors and family heads, members of the cultural majority and representatives of the many minority ethnic groups. This disparate assembly may have had a common goal in agricultural prosperity, but it was by no means united in its specific vision of an agrarian community. The interests and outlook of the majority prevailed. But there were places for others as well. Hired hands and members of unpopular ethnic groups had their own special economic roles to fill and social niches to occupy. As long as they did not disturb the delicate fabric of social relations and adopted the majority vision, they were welcomed in the developing agrarian community.

6

The Nature of Work

"There was spice with every day's work and hard rough work but with it all a lot of fun."[1]

In May, 1909, Fred Pringle hired on for a month with George Hoskins near Stettler, Alberta. Pringle spent the first day discing, a tiring and monotonous job. The next day found him "puttering around," basking in the "fine weather these days."[2] He was soon back to more serious work, sowing oats one day, drag-harrowing the next, and ploughing two days later.

Pringle did not record the countless small tasks he performed every day; he noted only those he deemed the major part of the day's work. Thus, in one week he "sawed wood all day," "plowed all day," "broke in one of the big mares," and "sorted potatoes." The next week, he was "cutting brush" (and taking time to "bath in a pond"), discing, and fanning grain on a rainy day. By the following week the weather was right for planting, and he ploughed, seeded, disced, and fought off mosquitoes, which were "getting bad."[3] By the time Pringle finished his month on Hoskins's farm, he had added to his list of work hauling hay, hauling coal, and repairing machinery. The picture that emerges from the accounts of the working experiences of prairie agricultural labourers is one of continual variety. It helps to explain why men chose this line of work.

But even as Pringle was finding diversity and challenge in his work, he was well aware that he was a waged worker. He responded to the circumstances of his labour pragmatically, carefully weighing working conditions against wages, seeking the balance that would suit him best as a waged member of the agricultural work force.

Farm Workers as Labour

Hired hands, like workers in other industries and in other places, attempted to improve the conditions of their life and labour. In doing so, they found their working experience firmly rooted in their class position. Even while they worked to leave the working class, farm workers shared with it fundamental goals. The expectations they brought to their position as wage labourers and the conditions they encountered in the West shaped the contours of their experience of class. That experience in turn influenced the way they reacted to their position. Their response was pragmatic.

In the prairie West, hired hands employed strategies common to those used by labour elsewhere but tailored to meet the specific circumstances of agriculture during the pioneering period. In contending with such factors as a dearth of capital, the seasonality of labour demand, and the organization of the industry, they pursued strategies to take advantage of the particular circumstances of time, place, and industry. Their aims were those of workers everywhere, ranging from the intangible rewards of job satisfaction and control of their own labour in the workplace to the practical necessities of improved living and working conditions and, above all, higher wages.

Wages for farm work were notoriously low. Outside the high-paying harvest season, agricultural labour was paid less than was labour in any other western industry. Canadian "agricultural wages at best reach to within about 10 per cent of unskilled industrial wages," reported the *International Labour Review*, "but much more often range somewhere about one-half such wages [and] often below one-half."[4] The reasons bore little relation to the value of production.

The organization of the industry into small-scale units of production, in which the state of an individual farmer's finances decreed how much he offered his help, kept wages down. Beginning farmers were often capital-poor, yet well-established and even prosperous farmers maintained the prevailing low wages. Cases of better-off farmers offering higher than usual wages were rare, although pioneer Anna Farion recalled that her father and his brother, while living in Dauphin, Manitoba, in 1897, "hoofed it all the way to Brandon where farmers were quite well-to-do and able to pay good wages."[5]

The international market for agricultural products also kept wages low. Farmers paid their hired help at least partly on the basis of estimates of their own income, but more on the basis of the local "going wage." If they could anticipate a good crop and a good price for grain, higher wages could pay off. However, since wages were decided at the beginning of the season and profits or losses were not realized until the end, farmers attempted to hold wages to the minimum that men would

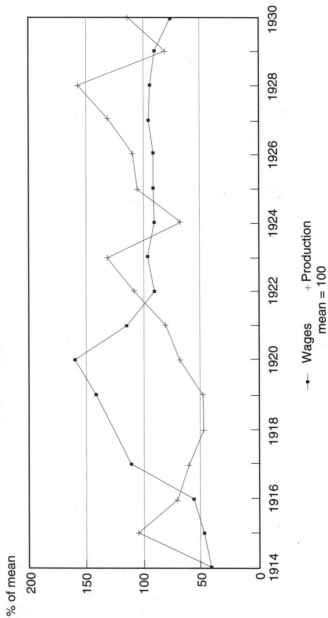

Figure 1
Wages and Wheat Production, 1914-1930

accept. Farmer George Tuxford of Moose Jaw, Saskatchewan, a former hired hand himself, complained in 1911 that "the labor problem is intense," with men "asking all kinds of wages. This might be borne with a good crop," he reasoned, "but under existing conditions are out of the question."[6] While the principle of shared risk was a familiar theme in wage negotiations, farm workers seldom received bonuses for a bumper crop or for unexpectedly higher prices. More often, hired hands were faced with employers who found at the last minute that they would not be able to pay the wages they had promised, or who were unwilling to do so. When Ebe Koeppen found threshing work to help finance his homestead, he had difficulty collecting his wages. His employer "wouldn't, or couldn't pay up. I had to ride over there three times in the middle of winter till I got most of my wages."[7]

But the single most important element to determine the low wages, or indeed the occasions of high wages, was simply the labour market. It almost always seemed as if farm labour was in short supply, especially during peak periods. In 1912, the Saskatchewan agriculture department estimated that 25,000 additional workers would be needed to harvest the crops. The province received 15,000 harvesters. The next year, the department estimated the need at 20,000 and received fewer than 13,000.[8] During the temporary peak labour demands of harvest, and in periods of acute labour scarcity such as the First World War, farm wages shot up.[9] When the labour supply was greater than the demand, as during the winter months and in periods of depression, farm wages plummeted.

Hired hands had to be both astute and fortunate to be able to tailor their wage requirements to the state of the labour market over the course of a year. If they were attempting to build a nest egg, the task was even more difficult. The promise that a man could earn enough as a hired hand to begin farming contained a germ of truth, but a number of considerations served both to reduce the wage tally and to make inroads on the accumulating bankroll.

Wages were not only low, they were extremely varied. There was no uniformity to the demand for labour. Farm jobs might be plentiful in some areas and scarce in others depending on the condition of the land or crops, the type of farming, and the size of the local labour supply. In Saskatchewan in 1914, a farm worker could earn as little as $230 per year including board in the lowest-paid North Western district or as much as $338 per year including board in the highest-paid South Central district.[10] The seasonal cycle of agriculture created enormous oscillations in labour demand, which were naturally reflected in wages. If a farmer needed help urgently he raised his offer, causing farm wages in his district to fluctuate, to the great dismay of his neighbours. For farm hands, the irregularity had as severe an effect. The newcomer who believed the assurances of some immigration propaganda that he could expect wages of $50 or $60 a month soon found that this might be available only for the spring and summer months or for the peak harvest season. Winter wages were seldom more than $5 a month, or could be only room and board and a bit of tobacco money.

Wages varied for other reasons as well. An examination of farm employment by the *Labour Gazette* in its first year of publication, 1901, found that variations in wages were "largely attributable to the variation in the efficiency of the different employees."[11] By the turn of the twentieth century, according to the Department of Agriculture of the North-West Territories, the prairies had absorbed most of the experienced farm workers that eastern Canada could supply. The result was inexperienced, inefficient, and therefore low-paid farm labour, especially at harvest. The department's *Report* painted a gloomy picture:

> The glowing possibilities of the West having already attracted there permanently a large proportion of the farmers' sons and those experienced in farm work from the eastern provinces, many of those who take advantage of the excursions are found to be unemployed men and boys from eastern cities who cannot satisfactorily perform the duties required of them on a farm without a certain period of training during which they are as a rule worth little more than their board.[12]

Men who possessed a wide range of farming skills found that their experience usually earned them a wage that could range from an extra $5 or $10 a month to more than double the wage of an inexperienced worker. The "better classes of men are preferred even at the higher wages," reported the *Labour Gazette*, "most farmers being prepared to pay to good, able-bodied, competent and trustworthy men the maximum wage quoted, rather than to employ inferior men at lower rates."[13] In 1912, the Saskatchewan government began a program to subsidize the travel costs of experienced farm workers from Great Britain to work for much higher wages than they had been getting at home. The Department of Agriculture explained:

> While Saskatchewan farmers suffer from a dearth of experienced farm labour there are in Great Britain hundreds of thoroughly experienced farm servants who are debarred from coming to this province simply because it is impossible for them to save sufficient to pay passage money from the scanty wages received.[14]

Newcomers realized that their shortcomings would be reflected in their wages. Even with his agricultural background, Gaston Giscard admitted: "I can't estimate what I earn since I'm completely ignorant about everything concerning field work." He was willing to offer his services for a trial period. "It isn't that I'm not willing to try," he declared. After a week, he decided to "take a chance on asking: Everything all right?" and was given an affirmative answer and a wage of $20 a month.[15] Yet experience was no guarantee of a higher wage. Newcomers to a district sometimes found that they had to accept a lower wage until they proved their worth. This might mean an entire month's work at a beginner's wages, which could make a significant difference in a short-term job of two or three months. In a long-term job it could prove disastrous. When George Shepherd and his father hired on at a farm near Brandon, Manitoba, in 1908, their inexperience led them to agree to a wage of only ten dollars a month for the entire summer, including harvest.[16]

The seasonality of agricultural work was probably the greatest variable in determining a hired hand's income, regulating wages, length of employment, and categories of farm work.[17] At the bottom of the wage scale was the winter job, which lasted from freeze-up until spring and often paid no more than room and board. A full-time job was one that lasted the full year. Wages were lowest in this category but were assured for the full year. More common and more highly paid were jobs that lasted for the summer or the "working year" of a farm, that is, for seven or eight months. Even more common and more highly paid were jobs for the busy season in the spring or fall, a period of one to three months. Finally, the shortest-term jobs were undertaken for specific tasks and were paid by the day. Wages here depended on the job and could be as little as $1 a day for rock-picking to $5 or $6 a day for harvesting. The latter was the most highly paid work but the most uncertain. Rain or a breakdown in machinery resulted in the loss of work that could completely offset the high wages. The "frequent rains so interfered with continuous employment on the farms that men quit work," reported the Saskatchewan Bureau of Labour in 1911. They were "unable to make a living wage owing to so much lost time."[18]

Hired hands responded to the challenge of low and irregular wages by developing strategies to take advantage of fluctuations in the labour market, turning a liability into an asset. The most effective strategy was the very simple one of withholding labour. Farmers were outraged by "the quasi-strike attitude" of farm workers who "show a tendency to stand out against reasonable wages."[19] According to all contemporary reports, the demand for labour was pressing and persistent throughout the settlement period. Farm journals regularly carried articles about the shortage of help, and farmers' organizations sought the aid of government and private labour and immigration agencies to fill their needs.[20] But labour shortages were inconsistent. During the pioneering period of the late nineteenth century and in areas just opening up in the twentieth century, the shortage of labour was absolute. But as regions passed beyond the days of sparse settlement and as the influx of settlers burgeoned after the turn of the century, the labour shortage became simply relative.

The seasonality of cereal monoculture further complicated labour demand and supply. During the harvest period of high wages, men flocked to the industry, yet farmers continued to complain that they were unable to find sufficient help to bring in their crops. The logistical problem of rapidly transporting the labour supply to areas of demand provides only a partial explanation. What economists refer to as lack of effective demand is the real key to understanding the existence of a sufficient labour force but an apparent labour shortage. When farmers were not offering employment, it was more often because of a shortage of cash than of a shortage of work to be done. Farmer Georgina Binnie-Clarke turned down two men who came looking for work after harvest, but the next day she listened to her brother, who was helping her farm, lament that "you simply can't get labour for love or money."[21] When the agricultural industry complained that it could not attract labour, its real difficulty was that it could not provide wages to compete with those in other industries.

The strategy of withholding labour was carried out at different levels. In the aggregate, it was reflected in the continuous refusal of the labour force to satisfy the labour requirements of the industry. But working men were also selective in their tactics. Clearly, it was to their advantage to ensure they were employed during high-wage periods. At the individual level farm workers employed tactics that demonstrated their ability to make most effective use of the shortages of labour, the seasonality of production, and the organization of the industry.

One of labour's basic aims is to maximize its earnings. Hired hands were sometimes torn between working to gain experience and working to build up a bankroll. When George Shepherd and his father took their low-paying job near Brandon, it was on the condition that by the end of the summer they "would become fully qualified to work on any farm in the West or to start up on our own."[22] But after six weeks, they balked at the low wages and left the job in search of something better. When men acted primarily as workers and secondarily as apprentices, they placed the priority of wages above that of gaining farming experience. Rather than settling down to a steady year-long job, they kept their employment options open. Wage figures for 1901 illustrate the point. Across the prairies, they range from as little as $10 per month for a yearly engagement to as much as $2.50 per day for a shorter engagement.[23] In the Lisgar district of Manitoba, average wages were $156 for a full-year engagement. A man could thus earn $13 a month if he hired on for a full year, but this increased to $20 a month for a seven- or eight-month job and went up to $25 if he hired on for only a month or two. Thus during periods of labour scarcity, when a man could find farm work quickly at the best rates, he could calculate that he could earn more by moving from job to job. A season's job lasting eight months could pay $160. He could earn even more by changing jobs more frequently. If he filled his year with a winter job paying only room and board, then a number of short-term jobs in the spring, then found a four-month summer engagement, and finally moved into the harvest fields in the fall, his earning potential rose to more than $200.[24] It was a gamble, but one that many were willing to take.

The effectiveness of the tactic depended on the use farm workers made of the particular conditions of prairie agriculture. Individual small farms were the backbone of the industry, and each was a potential source of employment. Farmers outnumbered hired hands, even though the number and proportion of farm workers increased until the beginning of the Second World War. But most farmers needed help at some time during the year. Farm workers used the large marketplace for their labour as a lever to pry up their wages. In conjunction with seasonality of production, the tactic could be very effective. Especially at harvest, men were able to parlay the urgency of the demand for their labour into higher wages. Although farmers sometimes made agreements among themselves to keep harvest wages at a uniform level, they undercut their own potential strength by engaging in bidding wars during the crucial weeks of late summer.[25] Farm workers were quick to take advantage of the competition for their labour, and they were equally quick to move on to another job if a better opportunity presented itself.

The Nature of Work

The individualistic strategies farm workers developed to deal with the problems of low wages proved equally effective when they used them to shape their jobs to their needs. Hired hands had a close relationship to their work. According to sociologist Howard Newby, farm workers occupy a world of their own:

> Work on the land has customarily been regarded as a qualitatively different experience from work in any other industry. The agricultural worker tends to think of himself as being part of a distinctive breed, divorced by temperament and environment from the urban, industrial majority. The actual work itself is therefore an exceedingly important attribute of the farm worker's personal and social identity – more than many other jobs it defines for the farm worker what he *is*.[26]

Farm folk reminisce fondly of the pattern of agricultural work, describing its cyclical attunement to nature. "We tread two endless wheels of labour on this farm," recalled pioneer Robert Collins, "one within the other: a daily round of chores spinning inside the greater circle of seasons' tasks."[27] Farm work was arduous and often monotonous. It could be stultifying and unrewarding, yet it held an appeal that defied conventional explanations of economic returns for labour expended. In part, this was because the agricultural community lived on hope, on next year's crop, in "next-year country."[28] But beyond this stubborn optimism lay an ambivalence toward the work itself. Farm labour was demanding and tedious, yet it was also diverse and challenging. It offered such a wide range of tasks that even though boredom took its toll, the order was more often hustle and intensity. The backbreaking and muscle-stretching aspects of the work that could result in physical debilitation could also lead to physical well-being.

Perhaps the strongest appeal to the labour of farm work was ideological. The challenge it presented was embraced as an affirmation of both moral and physical worth. For men who chose farming as a career, the hard work, the diversity of labour, and the pressure of balancing time and effort and money might be frustrating and wearing. But it could also add up to an immediate sense of accomplishment and a deeply felt sense of fulfilment. Looking back on his years as a hired hand and homesteader, Ebe Koeppen recalled the pleasures. "Through thick and thin, I always found farming very rewarding." He "discovered that taking care of animals and working with growing things gave me great satisfaction."[29] Farm work could be a reward in itself.

Hired hands sought a share of these rewards. Men who entered the waged agricultural labour force did so by choice, motivated as much by the non-remunerative aspects of the work as by their prospects for the future. The low wages were scarcely sufficient compensation for the long hours, the strenuous labour, and the sometimes dangerous conditions.[30] There were easier and more secure ways of earning a living. Indeed, men whose primary goal was an agricultural future often turned to work in other industries to tide them over during periods of unemployment in agriculture, or to provide them with the capital to set up farming, or to subsidize their farming ventures. Murchie's study of the agricultural ladder found that significant numbers of farmers had engaged not only in agricultural labour but also in some other occupation before establishing their own farms. At Kindersley, Saskatchewan, ninety-four out of 198 farmers had worked at non-agricultural labour for an average of 5.8 years and at farm labour for an average of 6.5 years before becoming farmers themselves. At Turtleford, Saskatchewan, eighty-seven out of 178 farmers had held non-farming jobs and at Olds, Alberta, eighty-six out of 108 had done so.

The men engaged in a wide variety of non-agricultural occupations, including general labour, skilled artisanship, clerical work, civil service, commercial employment, entrepreneurship, and teaching.[31]

Yet they returned to farming when they could, finding in the labour itself present satisfactions and potent hopes for the future. Even more than a higher income, farm workers sought satisfaction in their working lives. Despite the hard work, Edward ffolkes insisted that "these two months of simple labouring life, like a plough-boy in England, will have been amongst the happiest and most undisturbed in my nineteen and a half years' residence in the world."[32]

Labour in agriculture is varied and complex. Each task requires a different degree of expertise and range of skills, and each brings a different degree of pleasure or misery. Farm hands have left few records of their private lives, but have been more willing to record, sometimes in detail, sometimes more cryptically, their experiences of the labour that filled their lives. Noel Copping, who began working as a hired hand near Earl Grey, Saskatchewan, in April, 1909, described the daily round of work on a grain farm:

> At present our daily routine of work is 5 A.M. to 5:30 P.M. Rise, milk cows, feed and clean down horses. . . . About 6 A.M. have a wash and breakfast. After breakfast I saw wood for the kitchen stove and get water from the well. . . . Then at about 7 A.M. we commence work on the land. . . . This morning I have been ploughing . . . at 12 noon we come in to dinner, first unharnessing the horses and putting them in the pasture. After dinner I clean out the stable then bring up the horses, feed and harness them. Then work goes on again until 6 P.M. At this hour we come in from the fields, unharness the horses and give them oats. Tea is the next item on the programme and afterwards the horses are turned out and the cows brought up to the stable and milked. This I usually do in the evening. Then any odd jobs are done and the day's work is over. I usually end up with a wash and am then ready for bed.[33]

Other kinds of work were called for in the yearly round. The "work on the land" to which Copping referred varied from season to season.[34] Spring was the time for seeding, from about mid-April to mid-June. Ploughing was the first order, when the soil was opened up and broken. The techniques and labour for this task varied widely. The simplest was turning over one furrow at a time by a single-blade plough drawn by a single horse or ox. A man followed the plough, working hard to cut a straight furrow and to ensure that the soil was turned over properly to expose the undersoil. The process could be speeded up using a gang plough, a series of two or more blades lined up to cultivate more land with each turn around the field. The extra blades added the

A hired hand who mastered a team and gang plough was highly valued. (Saskatchewan Archives Board, R-A111.)

complication of a two- or four-horse team. Different types of soil required different handling. It took strength and skill to cut a furrow two inches deep into virgin prairie sod. When farm communities held fairs and exhibitions, ploughing contests were high on the list of favourite attractions. The newly turned sod then had to be backset, or ploughed between the furrows to turn the undersoil to the top.[35] Stubble land needed less handling, but no less finesse, and a straight furrow was the mark of an accomplished ploughman. Hired hand Arthur Jan ruefully admitted his initial difficulty. "I would start off all right to plow a field with four horses but where I ended up was doubtful."[36]

Ploughing was only the first step in preparing the soil. Next came discing, further breaking up the soil by a series of discs pulled behind horses and guided by a driver. After the discing came harrowing, smoothing out the soil by pulling a long rake-like implement over it. Hired hand John Cowell commented about his turn at the task. "At Sniders my job was harrowing with 4 horses. Walking. My feet was sore before I started."[37] Then the soil was ready for seeding, accomplished by pulling a seed-drill press over the land, with levers to regulate the depth of seed. Finally, the land was harrowed yet again, to ensure that the seed was properly covered. All these tasks had to be completed in a rush, with an eye on the weather. The time between spring thaw and first fall frost was measured in days. A week's or even a few days' delay could mean the loss of the crop.

Once the crops were planted, the summer lasted until early August. There was always work to be done, as a popular quip from Weyburn, Saskatchewan, attests: "You can rest while you feed the pigs."[38] Farm lands were maintained and improved. Summer was often the time to break new land for next year's crops. The simple turning of prairie sod was only possible on stretches of fields that were comparatively flat. Most farms contained brushy areas that needed clearing. In the parklands, trees had to be removed as well, and it was a rare prairie farm that did not have its share of rocks to be picked. Dry-land techniques dictated ploughing summer-fallowed fields and, depending on the soil, discing and harrowing as well. This was also the time for general farm maintenance and improvement. Wells were dug, farm buildings erected or repaired, fences strung, and machinery and implements overhauled. As the summer drew to a close, haying began, and then preparations were made for the climax of grain farming – the harvest.

The rush and labour of harvest far surpassed that of all other seasons. The vast acreage to be harvested required an enormous amount of manpower, and the threat of fall frosts dictated that preparing the crop for threshing be done very rapidly. Technology speeded up parts of the process, thus increasing labour needs. The first step in harvesting was cutting the crop. Teams of horses pulled binders into the ripe grain fields. The binder contained a moving blade that cut the standing grain a few inches above its roots. The stalks fell onto a moving belt and were carried to a knotting device that tied them into sheaves, which were dropped onto the stubble.

Manpower took over the process at this point, and two or three men followed a binder to stack the sheaves into stooks. Although stooking was "a muscle aching piece of essential hand work,"[39] it required more finesse than outsiders imagined. "You take two [sheaves], put them down and then put eight around in a circle," recalled harvester Philip Golumbia. "You would put them in two and put a little on the side to let in the air so they wouldn't shrink. They had to be done just so."[40] The sheaves had to be stacked skilfully, each leaning inward at just the right angle to maintain balance against winds, yet with all grain heads exposed to the drying air. The prospect could daunt even seasoned farm workers. When Bernard Harmstone, a hired hand from Quebec, took a harvest excursion to Caron, Saskatchewan, he recalled that he "looked at the . . . bundles [waiting to be stooked] which lay in rows as far as the eye could see & my heart sank down into my boots." After stooking for a while, he was dismayed to turn around and discover that "the stooks which had been so laboriously built were all laying on the ground." But he kept at it, and "with a little more practice I caught on to the trick." The job left an indelible etch. "For the rest of my life the memory of all those bundles laying there in rows all the way to the horizon & waiting for me to come along & pick them up, well words really can't express the feeling."[41]

The next step, threshing the crop, separating the kernels of grain from the stalks, called for another type of technology and another infusion of labour. On commercial farms during the pioneer and settlement period, huge steam engines powered the separator that threshed the grain.[42] On an average farm, eighteen to twenty men fed the grain and tended the machines. An engineman and waterman supplied fuel and water to keep the steam engine running. Other men kept the separator fed. In the fields, men dismantled the stooks, pitching the sheaves onto horse-drawn bundle wagons to be taken to the threshing machine. The key men were the field pitcher and the spike pitcher. "A spike pitcher had to be a tough, strong and steady man and he had to work ten to fourteen hours a day," according to harvester S.J. Ferns.[43] The field pitcher helped load the wagons, ensuring that the sheaves were evenly distributed in the wagon. At the separator, the spike pitcher took over, climbing onto the wagon and pitching the sheaves into the feeder. Care and efficiency were needed to ensure that the sheaves were placed correctly for the knife to cut the binding twine and to spread the stalks for the header to cut off the heads of grain. The grain fell to a series of rapidly moving tables inside the thresher, which separated the grain

Large threshing crews called for teamwork. (Provincial Archives of Alberta, B273.)

from the straw and chaff. Fans blew the straw and chaff out of the machine through a huge galvanized sheet-iron tube, and another metal tube delivered the threshed grain to a wagon or to sacks to be sewn up and loaded for market.

Other fall jobs seemed leisurely and secluded by comparison. Hauling the grain to market was solitary work, taking as much as a full day for a single load if the farm was some distance from the railroad. Farm hand Edward Corcoran enjoyed the job. "One always appreciated the nice restful morning when one was sent to Weldon [Saskatchewan] with a load of hay, . . . as it meant a comfortable journey on a nice sunny day with nothing to do except smoke a pipe and think philosophic thoughts."[44] Fall ploughing or harrowing was not under the same time pressure as that done in the spring, and field work gradually slowed with the fall frosts and the onset of winter.

Winter work was deemed even more leisurely, but it consisted of a lengthy round of regular chores. Percy Maxwell dispelled the notion that winter was a time of "enforced idleness":

> Chores take up nearly all day now that most of the cattle are in the stable. . . . There are 15 horses and a cow and three calves in my stable and it takes more than five minutes to clean and feed that little crowd. Besides the chores there is hay and wood and straw to be drawn from the bush and stacks respectively.[45]

All jobs were made twice as time-consuming and onerous with frozen wells and blizzards, as Maxwell's lengthy description of the relatively simple task of "drawing wood" illustrates:

> It is not a pleasant job. We get up in the middle of the night and do the chores then have breakfast in as big a hurry as if we were going for the 8:34 train instead of the 9:4 [sic] and start off at daylight. Arrived at the bush we cut down the trees and load them on to the sleighs and immediately start back for home and just manage to get in before dark. There is no time to stop for dinner and we have to munch a few bits of bread and butter (generally frozen solid) sitting on the load and even that is at the risk of upsetting because the roads are very bad and if we don't give all our attention to driving, the load will probably turn over. We have to cross the creek three times and there are exciting moments rushing down one bank on to the ice and up the other bank again; it is a tricky business too driving round the stumps and trees, and by the time we reach home we are pretty well tired of the 22 mile drive, especially when a snow storm comes on and your face is covered with a kind of wet ice and eyelids freeze down whenever you blink.[46]

Winter was also the time to begin preparations for next year's crop.

There was no rest, as hired hand Ray Coates recalled, for "as soon as dinner was swallowed the boss got an uneasy look on his face and . . . we were hustled out to bluestone wheat."[47] Seed had to be fanned and selected against weeds, and implements and machinery repaired.

The lengthy roster of farm tasks described here only scratched the surface of work on a prairie farm. The experience of hired hands encompassed much more, since they were expected to turn their hands to any task that might turn up. They cleaned stables, milked cows, collected eggs, butchered pigs, dug wells, hauled manure, and pulled sow thistle. The list was endless.

The variety of the work does much to explain why men chose this low-paying, arduous labour. Edward ffolkes described to his mother his "decided interest" in farm work: "there is at least one feature in it which you know I like – change, a great variety of things to do and attend to."[48] Rural sociologists have long recognized that the "irksomeness of farm work . . . is greatly relieved by its variety . . . [which] lessens the monotony of toil and affords rest thru [sic] change."[49] Farm workers might spend hours and even days at the same work, but they interspersed it with different jobs, and sooner or later were compelled by the inexorable agricultural cycle to move on to other tasks. Hired hand W.N. Rolfe of Manitou, Manitoba, found farm work a "continual round of exciting adventure."[50]

The alienation that beset workers in most mass-production industries was largely absent in agriculture.[51] The variety of agricultural tasks meant that in their daily and yearly round, hired hands were called on to employ a wide range of skills. They made discretionary judgements and had to adapt to a varied and changing environment. Their labour processes were not fundamentally directed by agricultural technology, and they were closely and tangibly bound up in the processes of production. Percy Maxwell remarked on the difference between "the useless sort of existence" at his previous employment and his work on a farm. "There is something to show for your work every day, so many acres of land ploughed, etc. and not simply one more day over and wasted as it was in the office."[52] Hired hands were able to see immediately the fruits of their labour.

There were many rewards. The ability to learn and exercise new skills provided a satisfying sense of accomplishment. George Shepherd and his father were proud of their growing knowledge. "We were no longer 'green Englishmen,'" bragged young Shepherd, "but trusted hired men."[53] Men also found satisfaction in the physical demands. "I am doing first rate," wrote John Stokoe to his father, "thriving like a young bullock, growing out of my clothes & enjoying myself immensely."[54] In later years, farm hands could be nostalgic about the hard work. Recalling his 1905 harvest experience, S.J. Ferns declared that "there was a real joy in feeling one could do a man's job in the sunshine and the wind in Manitoba."[55]

And since agriculture provided more than mere livelihood, hired hands could be captivated by the more esoteric aspects of the agrarian life. Noel Copping was moved by the beauty of his physical environment. In his description of breaking the virgin prairie, he digressed to record the vanishing scene. "Little yellow violets grew among the prairie grass, and in some spots there are minute strawberries sweet to the taste." As he followed the plough after his boss, kicking down improperly fallen sods, he had watched "a pretty blue bird hover, reaching for insects in the sod turned up by the plough, its note like liquid poured from a bottle."[56]

But the aspect of agricultural labour that held the greatest appeal was independence. Men were drawn to the West for the independent future it offered. They expected to work independently in building their own farms, and as a prelude they sought the type of waged labour that offered the highest degree of on-the-job independence. Hired men put in long, hard days, yet they were seldom kept under an employer's watchful eye. They sought autonomy and found it.

The men who were drawn to agricultural labour thus worked under conditions that suited themselves as much as their employers. Self-sufficiency of labour was a necessary requirement in an industry that suffered a chronic labour shortage and in which capital also performed the bulk of the labour. The most valued qualification of a hired hand was independence, the ability to work at a complex array of farm tasks without supervision or instruction. Even a newcomer was required to carry on unsupervised. Gaston Giscard began his "apprenticeship in ploughing" with the boss at his side. But after "one or two rounds to show me how to operate the levers, . . . he leaves me, telling me to go on doing the same for the rest of the day."[57]

But from the perspective of labour-capital relations, independence was a double-edged sword. Capital sought to harness this independence to produce a work force that was self-sufficient in its abilities yet dependent in its need for waged labour. Labour responded by building its strategies of resistance on the foundation of independence.

The promise of independent ownership ensured that farm workers did not challenge the basic nature of labour-capital relations. Yet for all they may have found general satisfaction in their work, hired hands found fault with many aspects of it: the uneven and onerous conditions of work, the poor living conditions, the long hours, and the low pay. But rather than interpret their grievances in terms of a fundamental conflict between their aims and those of capital, they regarded them as merely occasional or specific complaints against the conditions of a particular job or the demands of a particular employer. When they did object, it was to specific problems.

Like workers in other industries, hired hands tried to soften the harsh realities of their experience of class. They sought to exercise control over the pace of work, to avoid the more burdensome tasks, and to resist the demands of exacting employers. They gained a measure of control over their conditions of work by employing strategies that were highly individualistic. With the crying shortage of agricultural labour they could easily find other work under better conditions, whether their criteria for "better" included higher wages, or better food, or greater opportunities to learn. Their criteria depended greatly on their individual circumstances and aspirations, and could shift dramatically in the course of their work cycle.

Men set their own limits to what conditions they would tolerate. At times they were willing to accept low wages, at other times poor conditions. When Fred Pringle found work on a threshing crew, he remarked that "Men mad as wildcats" because "Beds not in." By the next day he was "Mad as a wildcat" himself. Although the men let their anger be known, they were willing to put up with the conditions because wages were high. Pringle made at least $73 and felt that the inconvenience was worth it. But Pringle's tolerance was limited, and before the job was finished he reported that he "Got fired because I wouldn't work in the night."[58] At this point Pringle had the option of finding work with another threshing crew, or hiring on with a farmer until freeze-up, or moving out to his homestead. He chose the latter. Pringle was acting in his capacity first as a farm labourer and then as a farmer.

Thus the strategies men developed in their position as apprentice farmers dovetailed neatly with those they pursued in their position in the agricultural work force. In the short term, the success of these tactics led to their continued use. In an industry dominated by a large number of employers, with abundant opportunities for at least short-term employment, and under constant pressure of time from the seasonal nature of its production, farm workers were able to respond quickly and most effectively to rapid fluctuations in demands for their labour by acting autonomously.

They drew on a large arsenal of tactics of resistance, manipulating their conditions of work in subtle ways to provide temporary relief. When Ebe Koeppen was put to work pulling stinkweeds, he declared it "the most senseless thing you ever seen [sic] in your life. . . . Absolutely insane." He "worked faithfully for a while" but was soon fed up, and "Just layed [sic] down in the field and snored away."[59] Other tactics required sensitivity to personal relationships. Harry Baldwin worked out an understanding with his employer in which "swearing was the staple of our conversation." Each morning, Baldwin would "prepare for the duel," which gave tacit recognition to the fundamental antagonism between himself and his boss over the amount of work to be performed, but which at the same time recognized the co-operative and egalitarian nature of the labour: "On such terms we worked well together."[60]

The common feature of these tactics was their autonomy. They could best be practised individually, with farm workers taking immediate advantage of a situation in which they could exercise control. The ultimate tactic was simple and direct: if they were dissatisfied with their wages or conditions of work, hired hands simply walked off the job. Gaston Giscard, farm worker, was unrepentant about leaving his employer to find new job experiences. But Gaston Giscard, established farmer, took a different attitude when he brought out a hired man from his native France, advanced the man's fare, and paid him the going rate, only to have the hand move on, in "a not-very-considerate manner . . . at ten o'clock that night without even unharnessing the horses." To Giscard's chagrin, he learned the next day that his man had "met a settler on the way and made an agreement to go and work for him."[61]

Job-jumping was such a common feature of farm work that men who moved from job to job were called "boomers" or "thirty-day men." Job-jumping was more than a simple inconvenience to farmers. It was a pragmatic response to an industry characterized by inconsistency and rapid flux in its labour demands, and it was a potent weapon in both the defensive and offensive arsenal of agricultural labourers. It was also one that was used widely by many other segments of the Canadian labour force, particularly workers in other resource and extractive industries, in construction, and even in mass-production industries.[62] It disciplined employers, influenced wages and working conditions, and shaped the contours of labour-capital relations.

Withholding labour is one of the most powerful weapons of collective labour action, yet it can be used effectively by individuals as well. Hired hands employed it as individuals. They had to be able to respond quickly to rapid changes in labour demands, and individual mobility emerged as the most adaptable tactic. It is not surprising to find farm workers employing a strategic package that was strongly characterized by short-term pragmatism and individualistic responses. They demonstrated the independence that was both cause and effect of their class experience, yet demonstrated, too, that class determined their experience.

Farmers could do little but rail against this independence. Georgina Binnie-Clark and her neighbour, Si Booth, shared the common feeling that for a hired man to leave his job abruptly was the "meanest trick I've known any man play."[63] But the time was approaching when labour practised the tactic so widely that it began seriously to hamper production. The agricultural industry would be forced to reassess the nature of its reliance on such a recalcitrant work force.

This was perplexing. During the period of agricultural expansion, labour and capital reached an alliance that ensured enough benefits to make both feel well served. Farm hands dissatisfied with their position could find many avenues of escape. Those who chose to remain in waged work enjoyed a demand for their labour that gave them a high degree of independence. At the individual level, farmers were faced with independent men like Charles Drury, who admitted "I was a funny fellow – no odds where I was working." Drury used the labour situation of the wheat boom to pick and choose his jobs according to his own criteria. His sense of independence was enhanced when he could say "Go to it" in the morning and find another job in the afternoon.[64] Farmers who needed labour found they could rely on the availability of land to subsidize their labour costs, but they still had to be cautious about conditions of work and had to temper their demands to the objectives of their workers. The partnership that resulted was an accommodation of necessity, but it was no less real. Despite instances of antagonism and even the effective use by farm workers of tactics of resistance, relations between labour and capital in prairie agriculture during the early years of the wheat boom were characterized not by conflict but by co-operation.

Labour-Capital Relations in Wartime

The partnership between labour and capital continued even after the booming economy began to falter in 1913. It was a time of distress for hired hands and their employers alike, but they did not see the economic contraction as a portent of the fundamental alteration in the relationship between labour and capital. They shared a common concern for the fortunes of the agricultural industry, whatever their place in it, and they looked forward to a return of prosperity.

Ironically, the First World War was the catalyst in redirecting the course of labour-capital relations. Wartime prosperity was of a different nature than that of the period of settlement expansion. If the wheat boom had appeared to have happened overnight, the war boom was even more sudden. There was an urgency to this round of fortune, since the war was expected to be short-lived. With apparently unlimited markets and an anticipated increase in the price of wheat, farmers rushed to expand their acreage.

Farm workers also took advantage. When they could, they began farming on their own account, and more than 89,000 homesteads were taken up during the war years.[65] To encourage this movement, homestead regulations were relaxed. An order-in-council under the War

Measures Act provided for homesteaders to count farm employment "as a like period of residence in connection with their respective entries," provided they were engaged in "actual farm labour" and could provide "sworn evidence satisfactory to the Minister of the Interior."[66] High wages in wartime industries outside the prairies lured many, as did the high enlistment rate among young single men. Those remaining were catapulted into a strong bargaining position. For the first time, the inability or unwillingness of farmers to pay wages competitive with other industries began to take a serious toll. Without adequate farm help, farmers would not only have to temper their expansion but would have to face the possibility of curtailing their operations just when they could reap the greatest profits.

The partnership built upon mutual accommodation began to show signs of strain. Observers warned that the familiar and satisfying relationship between farmers and their hired hands might not survive the determination of each to profit from the fragile prosperity. "Unless the war speedily ends farm labor of the better sort will be scarce," warned the *Farmer's Advocate and Home Journal* in 1915. "It may be that the scarcity of labor will develop conditions that will revolutionize methods of employing and handling workmen on the farm."[67]

The newly articulated antagonism between labour and capital manifested itself in familiar ways. Labour continued to use its most effective strategy, withholding its services, to push up wages. Capital continued to rely on its own organizations and government support to provide a labour supply and hold wages down. Neither tactic was new, but the special circumstances of war added a new force to the demands.

Farm workers sought higher wages immediately. Despite farmers' resistance, wages began to move upward, but the rise did not at first appear dramatic. Monthly wages increased slowly in the first year of war. But by 1916 in Saskatchewan, farm workers' wages were more than 25 per cent above their 1914 level. Harvest wages also showed slow initial upward movement. From $2.50 per day in 1914, Saskatchewan harvest wages rose to $3 per day in 1916.[68]

But as the war dragged on, increasing recruitment and manufacturing employment continued to draw men from the prairies. As labour shortages became more acute, farm workers pressed their advantage. In 1917, both monthly and harvest wages shot upward. In Saskatchewan, annual wages went to $458 in 1917 from $215 the previous year. Harvest wages rose to an unheard-of $4 per day. By the next year, Saskatchewan annual wages were up to $549, more than three times the pre-war figure, and were matched by harvest wages of more than $4.50.[69]

Given the high wheat prices, farmers might well have been able to manage higher wages, but they first attempted to do without help.

Some important agricultural practices, such as proper fallow tillage or fall ploughing, were easy to neglect.[70] The greater difficulties arose at harvest time, when the demand for labour drove up wages. Farmers importuned the various levels of government to regulate both labour supply and harvest wages, complaining bitterly that they were "being held up by unscrupulous laborers."[71]

Government response was cautious. "I have read your remarks concerning the hired man with interest," was the reply of a Saskatchewan Department of Agriculture official to one such farmer's complaint, "and I think probably some of the difficulties you have encountered at this time would arise from the fact that high wages are being paid, and a laborer has a tendency to become independent under such conditions."[72]

Farm workers did demand higher wages. They were met by more concerted action on the part of their employers, a legacy of the co-operative and associational activities of farmers during the pioneering and settlement period. Acting through the organizations they had been building to protect their marketing interests, farmers tried to use their influence to force government action. In 1916 the Balcarres Grain Growers' Cooperative Association passed a resolution and forwarded it to the Saskatchewan Bureau of Labour: the membership objected to farm workers taking "advantage of the present European War, in order to force up the price of farm labour. We do not consider it a fair deal to farmers of Western Canada." They petitioned both the federal and provincial governments "to introduce a measure to regulate the wage question for farm labour during 1916 or until PEACE has been declared."[73]

Farm workers from central and eastern Europe bore the brunt of the attack, facing hostility more strident than that described in Chapter 5. Angry farmers targeted "the foreign element of our Country," demanding that wages be set at $350 per year for foreign farm workers. They were willing to allow the wages of Canadian men to rise above that figure, but insisted that men earning more than $350 "shall be called upon to contribute at least $5.00 per month to Red Cross or Patriotic funds."[74] Other farm organizations across the prairies followed suit, translating the wage demands of foreigners and recent immigrants as disloyalty. But their real desire was for a cheap and docile labour force.[75] The resolution from the municipality of North Cypress to Manitoba Premier T.C. Norris was explicit:

> That this council go on record as being opposed to the exorbitant wages demanded by the aliens throughout the province and would suggest that $3.00 per day or $65.00 per month be the highest wages paid for harvest laborers and further we consider any alien asking higher wages should be interned.[76]

Such drastic proposals were not enacted, but immigrants expecting higher wages were faced with severe measures. When fourteen Austrians near Brandon demanded $4 per day instead of the $3.50 that local farmers considered reasonable, they were charged under the Alien Enemies Act. After getting off with only a fine, they were willing to work for $3.25 per day.[77]

Even more serious for farmers than the question of wages was the supply of labour, and it was here that the increasing tensions between labour and capital were most visible. Farmers felt beleaguered by the new-found strength of farm workers and used their organizations to urge governments to make "the most strenuous efforts" to provide farm labour. The Manitoba Grain Growers' Association petitioned the Union government to close "all non-essential businesses and [draft] men for farm work who are not engaged in essential operations," to return immediately to farms all "bona fide farmers and farm labourers who have been called under the Military Service Act," and to fix a maximum wage for "competent men and a minimum for boys and inexperienced men."[78]

Governments took no official action to regulate wages, but they did continue to work with the agricultural industry to find an adequate supply of labour. There would be no compulsion. "I am unable to see in what way the Government can interfere and insist on laborers accepting any particular pay," declared Manitoba's Minister of Agriculture and Immigration, "as we have no power to make anyone work who does not want to." He expressed regret at government powerlessness. "I know the situation is serious, but we are unable to do anything that I can see, I wish we could."[79]

But if governments could not force men to accept the farmers' conditions, there were indirect ways of providing a more pliant work force. The agricultural industry was able to enlist government aid in finding new sources of labour in groups more susceptible to control, such as soldiers-in-training. In Saskatchewan in 1916, the Bureau of Labour reported that "the Militia Department at Ottawa granted a furlough of thirty days to any non-commissioned officer or man who desired to accept work on a farm," and about 1,500 soldiers were granted leave. The following spring a Leave of Absence Board was established to grant leave to "any draftee who was a *bona fide* farmer and whose services were urgently required on the land."[80] In Alberta in 1916, "a little over one hundred hand-picked men from Vancouver, late in the season, [were selected] to fill the places of soldiers who returned early for camp duty."[81] At the beginning of the 1916 harvest, hired hand J.B. Burgeson wrote to farmer W.J. Blair for "work for myself and two of my friends[,] Carl Casparson and J.B. Larsen." The men were in training at Camp Hughes, Manitoba, with the 223rd Overseas Battalion and would not

Wartime supervision of farm labour undercut labour's strategic advantage. (Provincial Archives of Manitoba, Camp Hughes 1, c.1915.)

be granted harvest leave "unless we can by letter show for whom we are going to work."[82]

Other sources were sought as well. In Saskatchewan, the Bureau of Labour charged local committees with obtaining the "kinds of labour not heretofore fully or regularly employed in farming operations such as boys, girls, women, retired farmers, elevator and implement men, etc." These were workers who could be expected to accept conditions without complaint. Teenage boys were recruited under the auspices of the Canada Food Board and the Bureau of Labour. Arrangements with the Department of Education allowed the boys to leave school for farm work without loss of school credits. Those who worked steadily for at least three months were rewarded with a bronze badge for their service as "Soldiers of the Soil."[83] In Alberta, special provisions allowed children to work on farms, with lowered grade requirements for "qualifying examinations in the high and public schools."[84] Members of patriotic organizations consisting of "able bodied volunteers . . . professional and business men, merchants, clerks and artizans" were also pressed into service.[85]

Such labour sources as these could be counted on to provide workers who would not object very strenuously to conditions or wages. But regular farm workers, and the harvesters who joined them, could present a more formidable foe. When local sources of labour proved insufficient to meet the wartime demand, governments and the agricultural industry insisted that the net be cast wider. This carried new hazards, for it was a tacit recognition that farm workers were operating with a new strength. This in turn elicited a powerful response.

The spectre of radical labour organizations haunted government and industry. For farm workers on the prairies this was especially serious, since efforts to organize them came only from those groups against which strong measures of control were aimed. The Industrial Workers of the World, characterized by the press as "that mysterious organization of restless anarchical unskilled labourers rapidly spreading throughout North America," bore the brunt of the attack.[86] The experience of the "Wobblies" on the Canadian prairies clearly demonstrates the range and strength of concerted government and police reaction, as the IWW laboured under the double burden of espousing radicalism and of threatening the major western Canadian industry when it was most vulnerable.

In 1916, faced with an acute wartime labour shortage, the federal government issued a call for additional harvest help from the United States. This was an extreme measure, in view of the recent Wobbly success in American wheatfields. There, aggressive organizing tactics had caused membership in the newly formed Agricultural Workers' Organization to burgeon to 18,000 in its first year of operation. With

116 chapters by 1916, the AWO became the largest unit of the IWW, financially and organizationally underwriting expansion into the lumber and mining industries.[87] Anxious though it was to secure American harvest help, the federal government was diligent in its campaign to curtail labour radicalism. Compared to the pre- and post-war movement of harvesters from the United States, relatively few men actually crossed the border during the war.[88] Nonetheless, immigration inspectors were instructed to "use great care and caution that no I.W.W. agitators under the guise of harvesters . . . attempt to enter the country."[89]

Organizers were undeterred. By October, 1916, *Industrial Solidarity* was able to report with satisfaction that there were already "quite a few Wobblys on the job."[90] The local press responded less favourably. "The I.W.W. . . . are causing some alarm in different parts of Southern Alberta," warned *The Albertan* in August, 1917. "Reports of activity come to the city from Vulcan where they have been endeavouring to tie up harvest operations."[91]

Official response was swift and stern. Fearing an invasion of "IWW agitators who will certainly create ferment and foment disturbances," bringing "dissension among the threshing gangs, causing strikes or even burning up outfits,"[92] federal immigration officials worked closely with dominion and provincial police to prevent their entry to the prairies. Deportation proceedings were "obviously insufficient and ineffective" in combatting the expected influx of Wobblies, according to J. Bruce Walker, the federal Commissioner of Immigration, who sought to bypass the usual arrest and trial procedures in order to "cope expeditiously with numbers of these persons."[93] In his internal correspondence, Walker admitted that "our legal action in these cases has not rested upon a very solid foundation."[94]

In the meantime, his department was "making the most stringent regulations and taking the utmost precautions to prevent any descent upon Canada from Dakota of the I.W.W.'s who are in strong force there."[95] Farm workers at border points were carefully scrutinized "as to their connection with the I.W.W. movement, and upon the least suspicion of being in sympathy, were turned back to Uncle Sam."[96] In addition, immigration inspectors were sent to the United States to work with local officials who were equally anxious to put a stop to IWW activities. Fear of harvest disruptions and sabotage prompted co-operation among law enforcement agencies on both sides of the border. In 1917, a Montana sheriff warned his counterpart in Fernie, British Columbia:

we have a gang of I.W.W.'s who are enlisting men, or swearing them to go to Canada to burn the crops and destroy property, and I am going to put the fly-cops on them tonight and may make a cleanup. I am putting you wise so you put the Canadian officers on the lookout, as I get it they are to go there to work.[97]

If the Wobblies managed to enter Canada, they were carefully watched. Under the direction of the federal Department of Immigration, close surveillance was maintained on one such farm worker employed near Champion, Alberta, in 1917. Philip Lintz, a naturalized American of German origin, was "a rabid socialist," according to his employer, "and strongly in favour of the I.W.W. organization."[98] Superintendent of Immigration W.D. Scott declared that "His name is in itself sufficient cause to pick him up for examination as a suspected enemy subject." Immigration officers were advised that "If he carries no proof of his United States naturalization, I would detain him until he gets proof." Scott was anxious to have Lintz deported, suggesting that an American interpretation of regulations dealing with enemy subjects "might be useful in a case like this," or that a possible irregularity in his entry to Canada would "enable you to deal with him."[99] When this proved unsuccessful, a Royal North-West Mounted Police sergeant was assigned to keep continuous watch on his mail, his movements, and his contacts.[100]

Continuously under the watchful eye of the police, IWW organizers in Canada could do little. Even before the 1917 crackdown, Wobblies were closely watched and harassed. "There have been a few arrests for being too patriotic to the cause," reported *Industrial Solidarity*, when threshers at Weyburn were jailed and fined because "they kicked on the grub they had for breakfast." Others near Regina were sentenced to ninety days in jail "for refusing to work after 6 o'clock at night."[101] Finally, in 1918, an order-in-council declared the IWW illegal; membership became subject to a penalty of from one to five years in jail.[102]

Other controls on the agricultural labour force were less direct but no less effective. During the conscription debate and the 1917 federal election campaign, farmers, their sons, and their hired help were promised exemptions in exchange for support of conscription. But just as planting was about to begin in 1918, the blanket exemptions were withdrawn. Provincial governments took up farmers' pleas for relief from the labour shortage. Saskatchewan struck a Special Committee on Farm Labour, which recommended that:

> all men drafted, or who will be drafted, under the Military Service Act who have been farming land, bona fide farm labourers, . . . be given leave of absence as soon as possible, such leave to continue while such men are actually engaged in farm work.[103]

The report made a number of other recommendations to draw on unlikely or reluctant sources of labour, including women and children (urging compulsion in the case of teenage boys), and proposed offering exemptions for married men with children who would agree to engage in farm labour. To add teeth to the plan, the report recommended registering all militarily eligible men "for the purpose of assigning them to farms should the necessity arise." Finally, it urged amendments to the Criminal Code "to make the vagrancy provisions more readily available to unemployed persons."[104]

The federal government did agree to grant exemptions for farm labour, but farmers in need of help and men who wished to undertake the work had to apply individually, calling on provincial governments to support their applications.[105] Men who were thus released from war service had been screened and were compelled by the terms of their exemption to remain in the service of the farmers who had requested their release. Leslie Hutchinson was a hired hand from Saskatchewan who found himself tied to his farm employment by the red tape of exemption regulations. His former employer, who had already secured a short-term exemption for him the previous year, requested another release. "Leslie Hutchinson [is] a young man who has always worked on a farm and is well experienced in the use of all farm machinery," explained his employer. "I would like to have this man to help me run my machinery, as I find it almost impossible to get help of this kind."[106] Hutchinson may have welcomed the release from military service, but he was forced to pay a heavy price: losing his independent bargaining power just at the time when it was the strongest.

Individually, farm workers and their employers continued to work out compromises to the problems created by their divergent interests. But in the composite, labour and capital were determined to wrest all they could from the short-term prosperity of the war. Labour continued to withhold its services and to demand high wages, while capital continued to expand production. Calling for government aid tipped the balance in favour of capital, and farm workers found that their extra bargaining strength was severely curtailed. Still, wages soared in the later years of the war and labour-short farmers became desperate.

In self-defence, the agricultural industry moved in a new direction: to replace labour with machinery. The starting point was the reduction of labour-intensive farming practices. The opportunity to profit from high prices and to manage farm operations despite a shortage of labour caused prairie farmers to increase their specialization in wheat and to abandon recommended agricultural practices such as summer-fallowing and ploughing under stubble. In terms of production, the short-term result was higher profits while the long-term result was decreased yields. In terms of labour needs, the short-term result was a further concentration of labour demand at harvest time while the long-term result was a new perspective on agricultural technology.

Throughout the pioneering and settlement period, agricultural technology had been geared to agricultural expansion. Improvements in machinery and implement design that were labelled as labour-saving did make more efficient use of labour, but the agricultural expansion they made possible resulted in an overall increase in labour needs. Especially in wheat growing, rates of agricultural production increased faster than the growth of the agricultural work force. By 1911, the total male agricultural work force had increased by more than six-and-a-half times from what it had been in 1891, while wheat acreage was close to ten times what it had been twenty years earlier.[107] The war-induced crisis of labour supply gave a new meaning to the term "labour-saving."

The steps toward replacement of the labour force with improved technology were slow and halting. There was little innovation in implement design, and such machinery as gasoline tractors was not widely adopted. As late as 1918, the federal government seemed still to be coming to terms with the fact that agriculture was suffering from a shortage of labour. The Department of Agriculture in that year warned that the supply of farm labour was "gradually but surely falling off, with wages just as surely and even more rapidly rising."[108] Farmers had finally come to realize that the shortage would not be solved by longstanding methods of recruitment and control. At its annual convention in 1918, the Manitoba Grain Growers went on record as recognizing that "the scarcity of competent farm help must be replaced by labor-saving machinery, in order that vast acres of fertile lands now lying idle may be brought into use." They petitioned the Union government "to place all farm machinery and implements required in the production of food-stuffs on the [duty] free list."[109] Such a costly measure was unpalatable to the government, although it did agree to remove the tariff on gasoline tractors valued at less than $1,400.[110]

Technology might be a better answer. Government agencies encouraged farmers to think of making mechanical and technological changes to increase crop production while reducing labour needs. "Work Less Do More" urged a circular from the Minister of Agriculture: "Save Time, Men and Money by using More Horses and Larger Machines." Farmers received a detailed list of minor technological improvements that would prove to be "highly profitable even in times of normal wages." Disc harrowing with a single disc and team of two horses would cost $1 in labour to cover an acre twice, but the same operation "and a much better job" could be performed for only 60 cents if a double disc and four horses were used. More importantly, *"One man does the work of two."* Even a minor improvement could result in a substantial saving of labour. "If a double disc harrow is not available," counselled the circular, "attach a single disc throwing the soil inward to one throwing it outward and put on four horses. *One man's time is saved.*" Most farm operations were susceptible to such labour economies. A modification in the hay rack, using end pieces right across the rack, changed the job of building the load from "a good man's job" to "boy's work." Adding a sheaf carrier to the binder added two acres to the amount of stooking a man could accomplish in a day.[111]

Such modifications and even minor additions to the store of farm equipment were not major technological breakthroughs, but they were a recognition that technology was the key not simply to agricultural expansion but to the reduction of labour needs. The wartime demands of labour pushed agricultural technology along a new course, appropriate to the new climate of labour-capital relations.

The end of the First World War marked the closing of the second stage of agricultural development in the prairie West. In just two decades, the agricultural industry had experienced rapid growth and development, prosperity and recession and prosperity again, and a transition in the nature of the relations between its work force and its employers. Rural prairie society had also been transformed, from a pioneer frontier to an organized and diverse agricultural community based on the wheat boom.

Hired hands had a significant role in this economic and social development. In a number of different ways, they both adapted to and helped direct change. As ambitious, individualistic men who hoped soon to become farmers themselves, they contributed to the growing agricultural economy and they took their place in the developing prairie society. As members of the working class, they developed strategies to take advantage of the demand for their labour and to resist the directives of capital.

The First World War was a watershed. It created conditions that drove a wedge between labour and capital in prairie agriculture. The fundamental antagonism muted by the early years of the wheat boom was articulated openly, as workers and farmers used whatever tactics necessary to reap greater benefits from the fragile wartime prosperity. As individuals, farm hands and their employers coexisted co-operatively; as labour and capital in the West's predominant industry, they entered the post-war world eyeing each other with suspicion.

PART III

CONSOLIDATION, 1918-1930

7

Proletarianization

"If you talked unions you would be ostracized."[1]

John Grossman left his native Germany in the mid-1920s, hoping to find work as a hired hand before starting a farm of his own. He joined thousands of other European immigrants ushered into Canada under the auspices of railway companies importing cheap labour and potential railway-land purchasers. His journey was closely directed, and he was "advised to leave the train at Jansen, Saskatchewan," where "somebody was supposed to ask for me."

His initiation to Canadian farm work was not auspicious. His employer put him to work pitching hay "from sun up to sun down." After one week, recalled Grossman, "I balked, because my hands had started to swell." His particular grievance was not the labour – "I took the work in my stride" – but his relations with his employer. Neither the farmer nor his grown son pitched hay. Instead, they borrowed a haystacker and with a team of four horses built the same size stack they expected of him. Nonetheless, Grossman continued to work hard, and proudly reported that one day he broke the handle of the hayfork "on account of a big fork full of hay," something that "had never happened on the farm before."

Grossman continued to put his muscle into his work, hauling stones that the farmer's son declared were too heavy to move. But instead of praise, his reward was to be assigned "work which no one else wanted to do." Soon his diligence earned him the title of "dumb and strong."

He found this particularly galling since his employer could neither read nor write and called on Grossman to help his daughters with their reading. This brought more derision, and Grossman heard himself referred to behind his back as "this dumb guy." As his stay on the farm lengthened, Grossman became more and more dissatisfied.

John Grossman came to the prairie West when agriculture was becoming firmly enmeshed in capitalist production. During the 1920s, farmers looked to agricultural consolidation to solve the problems of a cost-price squeeze and growing international competition. Closer to home, they sought more effective methods of employee management to maximize their labour dollar. Hired hands like Grossman faced the consequences. The harder he worked, the more "the farmer looked at me, his hired help, as if I was just anoth[er] work horse."

Yet farm workers faced obstacles to improving their position. When Grossman asked his employer for permission to go to an exhibition in Saskatoon, he was refused. As a newcomer, he had entered a wage agreement that was well below the going rate. His employer "was afraid I might find out about farm wages and never come back."[2]

During the 1920s, the interests of labour and capital in agriculture began to diverge, as the economic circumstances of the waged sector of the agricultural work force came more clearly to reflect its economic position. Hired hands faced significant changes in their lives and labour. Increasingly, their experiences and their reactions were refracted through the prism of class.

The 1920s: Agricultural Consolidation

John Grossman's arrival in the West coincided with a decade of uncertainty for prairie agriculture. Overall, there was growth. By the end of the 1920s, there were more farms, more land under cultivation, and more agricultural production, but compared to the boom years of the pre-war settlement period and the surge of the First World War, the increase was uneven.

Prairie agriculture emerged from the war with highs in production and prices, but at the cost of lower grain yields and a heavy burden of debt. European agricultural recovery as well as increased competition from other grain-producing countries caused grain prices to fall by 1920, then continue to drop, then fluctuate widely over the decade.

The post-war slump did not strike immediately. In the first year of peace, when the federal government removed its $2.21 ceiling, the price of wheat rose to $2.24 per bushel. The following year it fell slightly to $2.18, but then the price began a sharp decline. By 1923-24, it bottomed out at $1.07. The relatively low prices for wheat during the first half of the 1920s reflected a generally depressed state for agriculture. The second half of the decade was little better. Wheat prices rose to $1.69 in 1924-25, then slipped to an average of $1.38 for the remainder of the decade.[3]

In the immediate post-war period the industry actually contracted, as farmers abandoned unproductive areas. In Manitoba, the interlake district and the area west of Lake Manitoba suffered population losses. The most severely hit area of the prairies was the dry region of Palliser's Triangle in southeast Alberta and southwest Saskatchewan. Originally ranching territory, the area was opened to agricultural settlement at a favourable time in the climatic cycle, which coincided with high prices for wheat. But by the end of the war the area returned to its normal aridity. After five crop failures in a row, from 1917 to 1921, disillusioned farmers began to move off the land.[4] The CPR colonization office reported that large numbers of farmers in southern Alberta were "so discouraged that there is bound to be an exodus from certain sections which are hit the hardest."[5] Between 1921 and 1926, farmers in the Palliser's Triangle abandoned 55 per cent of their farms.[6]

In other areas, new farms were created, especially in the northern parts of Saskatchewan and Alberta and in the foothills of the Rocky Mountains. At the middle of the decade, settlers began to move into Alberta's Peace River country, the wooded areas of Cold Lake and Beaver River in northeastern Alberta and northwestern Saskatchewan, and the Torch and Carrot River areas in Saskatchewan.[7] Occupied land had increased more than 35 per cent since 1921, with almost all the increase taking place since 1926.[8]

Farming continued to attract a hard-working and ambitious population, but the financial rewards were very unevenly distributed. The surest route was expansion. Successful farmers enlarged their holdings by renting or buying the smaller farms of their unsuccessful neighbours. By contrast, struggling farmers operated their farms on a shoestring and were in debt to a range of creditors from mail-order stores and implement dealers to mortgage companies and banks. Farm obligations doubled and tripled in real terms with the collapse of prices after the war.[9] Even increased production gave no guarantee of an exit from the economic dilemma as farmers came face-to-face with the great problem of agriculture: high yields and large crops usually brought a reduction in prices. Individual farmers were part of a commercial economy over which they had little control.

Although agriculture as a whole was prospering, individual farmers were faced with day-by-day and year-by-year decisions as they were hailed out, dried out, or bottomed out of the market in years when good crops and high yields led to a glutted market and depressed prices. The pressing question was whether they could keep on farming until the next upward movement in prices coincided with adequate rainfall, affordable labour, and no major disasters. The answer lay in their ability to predict the uncontrollable elements in farming and to manipulate those over which they could exercise some control.

During the 1920s, they began to move more forcefully to establish control over their industry. In their attempt to influence the larger economic world in which they operated, they relied on their agricultural associations, built on co-operatives to establish pools, and created political parties. In their attempt to control costs on their own farms, they stretched the traditional limits of relationships with their workers and found a new purpose for agricultural technology. These initiatives had direct consequences for farm workers and, in some cases, were precipitated by a labour force facing changes of its own.

Proletarianization

Proletarianization came piecemeal to prairie agriculture. It was an uneven and ongoing process, unplanned, unheralded, and largely unremarked. In an industry based on a synthesis of labour and capital, there was no provision for a permanent waged labour force. Rather, small-scale agriculture assumed that capital would also be labour. Undercapitalized prairie farmers could act as a cohesive and rational unit by providing labour for one another and thus furthering not only their own economic interests but the interests of their industry as a whole. It was a neat package, but one that did not work as planned.

Economic necessity did result in the provision of the bulk of the labour force from within the agricultural community. But the availability of free or cheap land during the pioneering and settlement period weakened the ability of farmers to hold on to their hired help, and for the busiest season the labour force was simply too small. Harvest excursionists solved part of the problem. They were recruited explicitly as agricultural proletarians: they did not own the means of production and they sold their labour for wages. Their position in the agricultural economy was temporary, to provide supplementary labour during the harvest season. During the rest of the year, labour needs could not so easily be met by a work force that compliantly disappeared once its work was done.

A permanent waged sector of the agricultural work force did not appear by design or because of distressed economic circumstances. The American census of 1910 revealed a similar tendency toward a growing agricultural wage labour sector in the United States, fuelled by the downward mobility of marginal farmers forced to give up their farms and enter the waged labour force. Such a prospect was anathema to prairie farmers and central Canadian policy-makers alike, since it struck at the very heart of the ideal of agricultural self-sufficiency as the key to economic independence. That this ideal was a myth for most prairie farmers, both in Canada and south of the border, has been well documented.[10] Men hung onto their farms as long as they could manage to supplement their farm income with off-farm work. The fear of sinking permanently into wage labour caused them to move out of agriculture altogether when they realized they could not make their farms into viable economic units or when they finally fell off the debt treadmill.

The agricultural proletariat in the West was thus not made up of failed farmers but of ambitious men who aspired to farm ownership and were willing to work hard to attain it. Hired hands were not, as were so many workers in the mass-production industries, former craftsmen and independent commodity producers trapped in wage labour. They were not, if from a working-class or peasant background, content to remain in agricultural waged labour on a permanent basis. They were not, if from Canadian and American farm districts that had filled up, hired hands and farmers' sons who were willing to spend the rest of their lives in an economic dead end.

Agricultural wage labour thus did not represent a decline in fortunes, but rather an intended first step out of wage dependency. The criteria for proletarianization in the case of prairie agriculture must therefore extend beyond the simple evidence of a sizable permanent sector of the work force that is waged labour. Hired hands had always sold their labour for wages, but their ultimate goal was to remove themselves from the ranks of wage labour. For them, the most significant development in the continuous process of their proletarianization was the constriction of the possibility of farm ownership.

It became much less likely in the 1920s that farm workers would become farm owners. The simplest route, through homestead entry, was becoming much more difficult to travel. The best homestead lands had already been taken by 1908. Land still available was agriculturally marginal – dry land in the southern parts of Alberta and Saskatchewan that had already been claimed then abandoned, or in the brushy or wooded areas to the north, far from transportation and marketing links and supply centres.

The reservation of lands under various schemes to aid British settlement during the 1920s further reduced the amount of land available to farm workers already in Canada.[11] And making a homestead entry carried no guarantee of success. The attrition rate was high. Between 1870 and 1927, the number of cancellations stood at more than 41 per cent of entries. In Alberta, the figures for 1905 to 1930 indicate a cancellation rate of nearly 46 per cent.[12] For later years the figures are even higher. On the prairies between the peak year of homestead entries in 1911 and the onset of depression in 1930, the cancellation rate was

Figure 2
Prairie Homesteads, Entries and Calculations

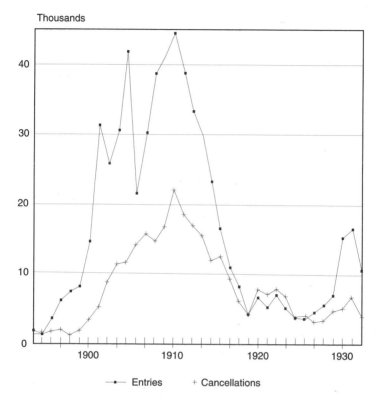

Thousands

—■— Entries   + Cancellations

about 60 per cent.[13] Homestead entries on the prairies declined steadily after 1911. During the 1920s the average number of entries was 6,394 per year, a far cry from the annual average of 34,638 in the decade leading up to the war.[14]

Purchasing land also became much more difficult in the 1920s as the combination of scarce homesteads and wartime inflation drove land prices to record highs. In 1912, the average price of Canadian Pacific Railway lands was $15.20 per acre.[15] By 1921, railway and Hudson's Bay Company land cost nearly $20 per acre, and land close to railways was fetching twice that amount. Good grain land during most of the 1920s cost $20 to $30 an acre.[16] Thus, the quarter-section grain farm that could have been obtained for a $10 filing fee before 1914 cost between $3,000 and $5,000 in the 1920s. Interest rates, which had already begun to move up by 1913, soared during and after the war to 8, 9, and 10 per cent.[17] In the 1920s the CPR shifted its land sales emphasis from a twenty-year payment plan to a thirty-four-year amortization plan.[18] Farmers who bought lands had trouble keeping them. Only in the last two years of the decade did the number of CPR land sales exceed the number of cancellations.[19]

In the meantime, technology had transformed prairie agriculture, making it necessary to purchase expensive equipment to bring large acreages under cultivation in order to pay off the cost of the land. In 1920, the *Nor'-West Farmer* pointed out the high cost of equipping a half-section farm:[20]

```
6 horses  . . . . . . . . . . . . . . . . . . . . . . . . . . .$1,200
1 wagon (complete)  . . . . . . . . . . . . . . . . . . . .160
1 set breeching harness  . . . . . . . . . . . . . . . . .80
2 sets general purpose harness . . . . . . . . . . . . .100
1 gang plow (12 or 14 in. with extra shares) . .150
6 section harrows or 4 section lever harrows  . .45
22-rim double disc drill  . . . . . . . . . . . . . . . . .265
8-foot binder . . . . . . . . . . . . . . . . . . . . . . . . .275
Mower and rake  . . . . . . . . . . . . . . . . . . . . . .160
Hay and sheaf rack  . . . . . . . . . . . . . . . . . . . . .25
Cultivator and packer (probably) . . . . . . . . . .400
        Total for equipment  . . . . . . . . . . . . . . .$2,860
```

As the decade progressed, machinery costs increased as farmers began to purchase tractors, trucks, and even combines. These developments drove the cost of farm ownership well beyond the reach of a farm hand earning the going rate of $425 or $450 a year during most of the 1920s.[21]

But immigration propaganda, official literature, and popular opinion refused to recognize that the position of agricultural labourers had become institutionalized. Contemporary observers clung to the conviction that farm work was merely a step toward farm ownership. There were always examples to support the claim that any hard-working man could become an independent farmer. Even in the depressed 1930s a hardy minority did establish self-sufficient farms. But during the 1920s such an achievement was expected to be commonplace. In 1926, the International Labour Office of the League of Nations asked the Canadian Council of Agriculture about agricultural labour in Canada. Secretary John Ward reported proudly that "The status of the hired agricultural worker in Canada is entirely different from that of the agricultural laborer in European countries." The work was transitory, he declared, whether seasonal or as a prelude to farm ownership. Even though a "very large percentage of the male population" was employed in farm labour at some time or another, Ward explained that "in almost every case such employment is engaged in with the idea that it will only be temporary." The result, he insisted, was that "In Canada we have practically no permanent agricultural laborer class."[22]

This view is reflected in current historical writing. Historians acknowledge that hired men occupied a disadvantaged economic and social position and that farm ownership was beyond their grasp, but they have been reluctant to define prairie farm workers as a proletariat. They are troubled by the transient nature of the agricultural labour force and by the shortage of the more explicit measurements of working-class consciousness.[23]

Hired hands have left little evidence of collective class action. There is no record of such recognizable characteristics of class consciousness as union membership or political activism. And even though farmers were often quick to accuse their employees of working against their interests, the men themselves have not left evidence of developing a comprehensive and effective campaign of class struggle. Farm workers did not seem to be collectively engaged in class warfare, and my own previous work reflected the historical consensus that they had "failed to develop a class ideology appropriate to their objective position."[24]

But how accurate is this assessment? Certainly, farm workers as a group remained outside the mainstream of union growth and radicalism that characterized western Canadian labour activity before 1920 and outside the political activity that typified its endeavours during the 1920s. In the historiography, farm workers are absent from the path of class formation, yet in the evidence they appear at every step along the way. Research uncovers many instances of their involvement in activity that is characteristic of class struggle, ranging from the very basic tactic of walking off the job to strikes and sabotage. Were these isolated incidents, or were they part of a larger and comprehensive response to capital? Capital and the state certainly recognized the class nature of these tactics and responded vigorously, attempting to control strictly the activities of union agitators among farm workers.

The attitudes of farm workers themselves are less clearly deciphered. Close study has failed to reveal a sustained critique of capitalism. They used specifically working-class tactics, but only sporadically demonstrated self-consciously collective class strategies. But increasingly throughout the 1920s, their lives were being refracted through the prism of class. Hired hands responded by acting in ways that reveal working-class approaches to their employers and to their status as an agricultural proletariat. The bottom line, in the words of Communist Party theorizer J.M. Clarke, was that "the nexus between [the farmer] and the proletarian he employs is that of the wage contract.

They must, and do, face each other as class enemies, as exploiter and exploited."[25] Few hired hands would have put it so starkly, yet by the 1920s the labour force was attempting to hold its own in a contest becoming more heavily weighted in favour of its opponent.

Barriers to Collective Action

Farm workers in the 1920s were at a crossroads. The strategies they had developed to deal with capital during the formative years of the agricultural economy were effective because they were individualistic. Hired hands had acted autonomously, using the availability of homesteads, the infancy of the industry, the shortage of labour, and the seasonality of production to reach satisfying bargains with their employers. But changing conditions called for new strategies. As free lands were taken, as the industry developed new tactics for dealing with its labour force, and as the labour shortage turned into a labour surplus, individualistic methods lost their effectiveness.

Yet farm workers did not easily switch their tactics. They were pragmatic, and found that individualistic strategies still remained the most effective method of exerting control over conditions of work. More significantly, a wide range of constraints reduced the possibility and potential effectiveness of collective action. Unified action required the agricultural proletariat to forge links with the wider labour movement. Even before moving on to this demanding stage, however, farm workers had to take the first step of uniting within their industry. This was a formidable task.[26]

Unlike workers in other industries, farm hands were widely and thinly dispersed. Small family farms relied as little as possible on hired labour, resulting in a ratio of paid agricultural labourers to farms in 1921 that was only one farm hand for every four prairie farms.[27] Even though the waged work force in agriculture grew by almost 41 per cent between 1911 and 1921 and by another 30 per cent during the 1920s, there were always more farmers than hands. During the course of the two decades, the ratio of farm workers to their employers went from 1:4.5 in 1911, to 1:3.9 in 1921, to 1:3.1 in 1931. And most farmers who employed permanent help kept only one hired hand.[28]

These logistical problems stumped farm hands and union organizers alike. "The farm workers are at the mercy of their masters," declared the One Big Union. "Being scattered, or in very small lots, over this vast country, they have no cohesion, no chance to meet together and safeguard their interests."[29] The great distances separating prairie farms, as well as the lack of leisure time, meant few farm workers could attend organizational meetings, as the Communist Party was acutely aware. "It hardly seems likely, does it," noted J.M. Clarke, that "the rich, exploiting agrarian is going to donate his car to the hired man to go to a union meeting at which wages, hours of labor, working conditions, etc. are to be discussed."[30] Organizers who hoped to engage in union activities while employed on a farm were equally hampered. The Department of Immigration might have saved itself the effort of trying to deport Philip Lintz during the war for his IWW activities. The Alberta Provincial Police noted with satisfaction that he was "working all day and had no opportunity of leaving the farm during the week to go out into the country or elsewhere to do any organizing."[31]

Even had the luckless Lintz been able to attempt to organize his fellow farm workers, he faced a deep division within their ranks. Seasonal workers who drifted into farm work during seeding and harvest were not a permanent part of the agricultural work force, and they were uncommitted to the long-term effort needed to improve their position within the industry. They were interested only in short-term goals, primarily higher wages. But long-term or full-time workers had to be cautious in their demands. Hired hands who had worked all summer or even all year on a single farm could not afford to jeopardize their accumulated wages for the chance of an extra dollar or two at harvest. When trainloads of harvest excursionists moved into the prairies each fall, regular farm workers welcomed the help and often the camaraderie of the big crews. But they were also wary. Hired hands occupied far too precarious a position to be able to make the same type of concerted demand for higher wages or better conditions that harvesters could and sometimes did. A comparison can be made between regular hired hands and, for example, the harvest excursionists of 1928.

The rich wheat fields of 1928 pointed to a bumper crop, and farmers began to worry early about the labour supply. The federal government responded by offering special harvest rates to unemployed British miners. They were promised steady work and high wages, and tantalized with the possibility of future farm ownership. But they found an overcrowded labour market and wages well below what they had expected. They began to turn down jobs, complain loudly to the press, and "soon found sympathizers with the Communist Party, . . . and [spent] the greater part of their time . . . in the Communist hall."[32] Regular farm workers may have sympathized with the miners, but they could not afford to join them. There were, as well, other reasons for the distance between the two groups. The British harvesters were almost entirely miners. They were a homogeneous group with similar backgrounds, language, and outlook, and they shared a strong working-class tradition. Their work in Canada was strictly short-term and temporary; they had nothing to lose by protesting low wages. Prairie farm workers, by contrast, came from a wide variety of occupational and ethnic backgrounds and had far too little time and opportunity to develop common traditions. Their need for full-time employment meant they could not afford the risk of antagonizing prospective employers.

Such a threat became more serious during the 1920s, as farm workers lost the labour market advantage they had enjoyed during the boom years. In the post-war decade, farmers continued the clamour for an ever-larger labour supply. But a substantial difference in the nature of the labour shortage went unrecognized in farmers' demands. During the pioneering period, the labour shortage had been absolute. The ready availability of land kept the labour force in continuous flux and in short supply. Even though many new farmers spent part of their year "working out" in order to establish their farms, the labour supply could not meet the demand. With the advance of settlement, and especially with continued immigration in excess of settlement possibilities, the labour force grew and labour needs could be more easily met. Still, the apparent problem of a labour shortage continued. The difference was largely one of definition. The shortage was becoming less absolute and more relative to the wages that farmers were able or willing to pay. By the 1920s, even though extra harvesters were needed for several weeks each year, the West had a labour surplus.

Ethnicity added a problematic dimension to the crowded labour market of the post-war decade, a difficulty exacerbated by the increased wartime hostilities discussed earlier. Prior to the First World War, newcomers to the prairie West had come in greatest proportions from eastern Canada, Great Britain, the United States, and northern Europe. But when immigration resumed in the aftermath of the war, a larger proportion of immigrants came from central and eastern Europe.[33] And after the middle of the 1920s, the proportion rose even higher, reflecting a mid-decade shift in immigration policy. During the 1920s, as before the war, federal guidelines for immigration aimed at encouraging bona fide agriculturalists and domestic servants. But in 1925 the federal government passed immigration selection into the hands of the railways.

Under the Railway Agreement, which remained in force until the end of the decade, the Canadian Pacific and the Canadian National Railways were allowed to determine for themselves whether or not prospective immigrants met the criteria. The railways received annual limits on the numbers of men who could be classified as agricultural labourers and were responsible for the costs of returning men who became public charges within a year of their arrival. Despite these restrictions, the numbers and proportions of immigrant men who entered as agricultural labourers were much higher than the numbers who actually engaged in farm work. They became a highly visible part of the general labour surplus that increased job competition and held down farm wages.

Not surprisingly, these immigrants faced serious difficulties in breaking into the farm labour market, facing opposition from farmers and fellow farm workers alike. Their experience was markedly different from that of ethnic workers in many other industries. In mining, forestry, construction, and even, in some cases, in domestic service, a strong ethnic identity and institutional network could provide a strong core around which organizing activity might coalesce.[34] Especially when the newcomers were overwhelmingly of the same class, groups with a strong tradition of collective labour action could provide the leadership and the impetus for organization. Farmers and other employers worried about such threats of organization, especially when they hinted of radicalism.

But European immigrants who entered prairie farming, although often from a common class background, saw themselves not so much as waged labourers, with benefits to be reaped from class organization, but as owners and potential employers. They were small-scale farmers, or aspiring farmers, and their common struggle was to adjust to new farming conditions and to establish themselves as farm owners. In these ambitions they shared an outlook that had much in common with their neighbours of different ethnic backgrounds. In the case of agriculture, ethnic identity tended to override that of class. Where the ties of ethnicity were strong, they tended to focus on common solutions to agricultural problems and on ways to strike a delicate balance that would allow them both to retain their ethnic identities and to carve out a social and economic niche. Although they were not always cohesive, they tended to divide themselves from other farmers rather than along class lines within their own ethnic group.

But when eastern and central Europeans sought work beyond their own communities, they met a cool reception. In the broad community of the rural West, the ethnic mix of the agricultural labour force served to divide workers. Since official policy limited the entry of central and eastern Europeans to those who were "bona fide agriculturalists," many who had neither a background nor an interest in agriculture were admitted as farm workers. Most of these were immediately drafted to other types of labour – in northern Alberta only 25 per cent of Ukrainians who had entered as farm labourers actually found farm work, with the remaining 75 per cent employed in construction and coal-mining.[35] Many, however, did enter the already crowded agricultural labour market. Immigrants were accused of cutting wages and of lowering the standards of working and living conditions. "It is extremely difficult," declared the Canadian Legion of the British Empire Service League, "for local citizens . . . to get work when foreign immigrants cut wages in half in order to get placed."[36] These accusations were not unfounded, since farmers regularly offered the more menial farm jobs and lower wages to eastern and central European farm hands than to those of British or northern European backgrounds.[37]

Reserving the worst agricultural jobs for immigrant workers was often defended in terms of their lack of experience or inability to speak the employers' language. But the practice was based on a more direct economic rationale. Much of Canada's immigration history is synonymous with the story of underpaid unskilled labour.[38] Farmers in the prairie West joined their voices to the demand for a surplus immigrant labour market. "I like to get cheap labor," explained one Saskatchewan farmer. "Naturally we can get these immigrants cheaper than we can get men who are raised in the country."[39] Although their lack of experience and language difficulties resulted in fewer jobs and lower wages, central and eastern Europeans did occasionally have an advantage over the equally inexperienced British immigrants, according to officials who noted that "foreign residents at country points will take a much greater interest in their compatriots than will English speaking residents in the British worker."[40]Job competition was exacerbated by intolerance. "Farmers wont [sic] you to work for $30 a month and know [sic] keep," stormed farm hand F.H. Dudley, "& if there is other work the Bor-Hunks [sic] get the first chance & the English men can starve."[41]

Flooding the labour market with immigrants created other obstacles to farm labour unity. The degree of agricultural expertise the newcomers possessed varied. Although the techniques of dry-land farming were still being learned in the early years of prairie settlement, prewar farm workers often came from agricultural backgrounds that enabled them to adapt to prairie conditions and to learn with their employers. Post-war immigrants were more likely to come from urban areas or regions of small-scale peasant agriculture. In the interim, prairie agriculture had been transformed into a highly technological enterprise to which adaptation was more difficult. The lack of appropriate agricultural experience helped to maintain distance between inexperienced newcomers and seasoned hands. Moreover, a surplus of inexperienced labour afforded employers the possibility of dispensing with lengthy periods of employment for experienced hands. They could be more cost-conscious in relegating tasks, allotting the unskilled work to the cheaper inexperienced men and saving the complicated and skilled tasks for the more expensive seasoned hands. In a labour market that was becoming more and more crowded, experienced workers could ill-afford the competition.

There were also more direct pressures against collective action. As the major western Canadian industry, agriculture could count on widespread support in controlling its work force. The legal position of farm workers was strictly circumscribed. Specifically excluded from ameliorative labour legislation, agricultural labourers were nonetheless bound by the stringent regulations of the Masters and Servants Acts.[42] The Acts made the carrying out of farm duties a legal matter and encroached on the personal lives of hired hands as well, legally enjoining them to sobriety and obedience, prohibiting "ill-behavior, drunkenness, refractory conduct or idleness," and forbidding them to leave the farm at any time, "day or night," without their employer's permission. Nor did the Acts provide any real protection. In the case of non-payment of wages, for example, farm workers could not collect more than two months' back pay, and this they had to do within three months. Since it was customary to withhold wages until after harvest, hired hands might work for the better part of a year only to discover that civil suits were their only recourse. But even these measures applied only to workers who had been "improperly dismissed." Farm hands who left their jobs before the agreed time were deemed to have violated their contracts, were not entitled to payment for any work already performed, and were subject to a fine or imprisonment.

There were other controls as well. The Saskatchewan Employment Service boasted that it had helped to mitigate "the spirit of unrest as evidenced by labour troubles the world over during 1920 and 1921." Had it not been operating, "the conditions of our farmers with respect to help would have been ten-fold worse than they actually were."[43] Employment agencies, immigration and colonization companies, and farmers' organizations, working in close co-operation with federal and provincial departments of immigration, agriculture, and labour, formed a strong organizational network. Their purpose was to provide a large, cheap, and tractable labour force. "I know you big farmers would like us to flood the cities and towns with men so that you could get them as cheaply as possible," wrote Saskatchewan's Minister of Labour in 1921, promising to try to secure "a number sufficient to do the work without creating a scarcity such as would encourage the boosting of wages."[44] Agricultural districts were canvassed to discover their requirements for both seasonal and full-time help, and immigration and employment agencies tailored the supply to the demand, effectively reducing the potential strength that might arise from a shortage of labour.

Newcomers to Canada were especially susceptible to the informal controls integral to immigration regulations. Entry to Canada or subsidized ocean passage was often conditional upon their undertaking farm employment; under the British settlement schemes of the 1920s, it was "absolutely essential."[45] In the early years of the 1920s, immigration from eastern and central Europe was restricted to bona fide agriculturalists and farm labourers. Yet even after the Railway Agreement, the federal Immigration Branch reserved the right to deport any who failed to find farm work or to settle on the land. Although this provision was not strictly enforced, threats of deportation or imprisonment hung constantly over the heads of these newcomers. In 1927, the One Big Union reported that in Winnipeg "destitute immigrants are sentenced to go to work on a farm for anything the farmer cares to offer, or be evicted from the immigration hall and then arrested for vagrancy."[46]

Follow-up services reinforced the control. Hired hands were encouraged to turn to placement agencies if problems with their employers arose. In such cases they were generally told simply to make the best of it. A CPR investigation of one such complaint revealed that the farm hand "seemed to be a nervous wreck and he stated that his position on [the] farm was unbearable." Nonetheless, he was urged, and finally convinced "after lengthy arguments," to remain on the job.[47] The Saskatchewan Department of Agriculture was even more direct. Its booklet of *Practical Pointers for Farm Hands* warned employees not to "dispute the plans of a boss, [or] demand things of the boss, [or] resent a harsh criticism. Don't try to bluff the boss by telling him where you can get higher wages," continued the advice, warning that "dissatisfied men are the first to be let go."[48]

Indirect methods of labour control were supplemented at times by stricter and more formal tactics. During the harvest period farm workers' strategic position was greatly enhanced; work stoppages or even delays could be crucial. But a careful monitoring of the harvest labour supply undercut this potential strength. The annual harvest excursions were carried out under the joint auspices of the railways and the federal and provincial governments. Distribution proved a major problem. There were always some areas that faced labour shortages for part of the season, but when workers used these opportunities to bargain for higher wages, they were quickly suppressed. "We have had some trouble with agitators," reported the Nanton Cooperative Association, "but when we found them we got the police to run them out of town."[49] Agitators might also face imprisonment. The Alberta Provincial Police reported that a number of harvesters who refused to work for the going wages "were brought into court and given the privilege of going to work or going to jail."[50] Many others were not given any choice but were simply jailed for vagrancy.[51]

The strictest controls were reserved for radical organizations. As the largest employer in the prairie West, and as the most important industry, agriculture had long been able to rely on the state to help control its work force. In the highly charged "Red Scare" atmosphere of the 1920s, federal and provincial governments were diligent in suppressing any organization with even a whiff of radicalism about it. Self-declared revolutionary organizations such as the IWW were prime targets, and its suppression during World War One continued in the following decade, reinforced by legislation arising from the war and the 1919 General Strike.

In 1918 an order-in-council declared radical organizations such as the IWW illegal, and membership in the organization became subject to a penalty of from one to five years in prison. The following year the Criminal Code was amended to cover illegal associations.[52] In agriculture, the prosecution of radical organizations could be carried out under a number of sections that defined crimes against the state in both general and specific terms. Section 134 of the Criminal Code declared "bolshevism" illegal, while sections 511 and 513 specifically singled out arson to barns and to wheat stacks. More informally, the RNWMP, and even the Alberta Provincial Police for about one year, were given authority over deportation and enforcement of passport regulations.[53]

Although much of the finances and energies of the IWW in the years following 1919 were taken up by legal proceedings, it continued its organizational work in the wheatfields. Its message was unequivocal. "To the farm hand the I.W.W. offers [a] modern labor union, one that knows how to fight the boss, a union that understands how to use the strike and the boycott."[54] In 1921, the IWW publicized a harvest strike in southern Manitoba, warning members to "take notice, and keep away" lest they be used as unwitting strikebreakers.[55] The RCMP had done their best to keep American Wobblies from entering Canada, responding to a warning that they were riding boxcars into the country. "Very close watch is being kept on the boundary for this class of person," reported the RCMP in an echo of similar diligence during the First World War, "and whenever they are located, they are summarily rejected and sent back to the United States."[56]

The efforts to keep Wobblies from Canada continued every harvest. In 1922, RCMP Inspector R.A. Gerrie and his officers were rewarded with a letter of commendation for their vigilance in keeping out the IWW who operated in North Dakota, a reputed "hot bed of I.W.W. bootleggers, whiskey runners and all kinds of criminals." Inspector Yardley was proud to report that he had "refused admission to 75 or 100, whom he considered were undesirables, and who, besides, carried I.W.W. cards."[57] The authority to keep Wobblies out of Canada rested on a broad interpretation of immigration regulations. The latitude that border patrols enjoyed during the war, when immigration official Thomas Gelley had admitted that "our legal action in these cases has not rested upon a very solid foundation,"[58] continued. In 1922, Gelley expressed hope that the IWW was an "organization of the class mentioned in 3(0) of the Immigration Act" and that prosecution for concealing a membership card when applying for entry could be allowed under section 33-2.[59] At deportation proceedings for notorious Wobbly organizer Sam Scarlett, the fact that he had declared his intention to find harvest work but had arrived in Canada too early in the season was introduced as testimony in an attempt to have him deported.[60]

In 1923, the federal Department of Immigration increased border surveillance when it received word from a "very reliable source" that "the I.W.W. element intend to pull a large strike in the harvest fields of North Dakota about August 20th and extend it into Alberta and Saskatchewan." IWW members might conceal their membership, warned the commissioner of immigration, "but will have the number of card written on inside of their hat or on a piece of paper or in a note book." More seriously, continued the commissioner, "it is also their intention to walk across the line and smuggle in cards and literature for the purpose of getting new members." Instructions were given that all authorities were to be notified, and "should any agitator be found among the harvesters he be severely dealt with."[61] The RCMP were kept busy trying to catch men who found the Portal, North Dakota, crossing "too 'Hot' for the I.W.W. element" and who resorted to jumping off the train that ran along the border.[62]

Some Wobblies slipped through the net. When "a few I.W.W. Agents" appeared in Weyburn, Saskatchewan, they were apprehended by the Saskatchewan Provincial Police and "were promptly dealt with . . . and returned to the other side."[63] Officials also found ways to deal with men who could not be expelled. In southern Saskatchewan, RCMP Constable R.F.V. Smyly received a warning that men were striking for higher wages in the harvest fields of Montana and might move northwest. His investigation uncovered men "working quietly amongst harvest hands near Shaunavon." He discovered that they were "trying to induce men to hold out for high wages, higher than those that are fair and reasonable." For the moment he was powerless, reporting that the "town Constable has tried to do something about these men but cannot 'Vag' them as they all have funds," but he assured his commanding officer in Weyburn that he would continue his investigation.[64]

Other prosecutions were more successful. When John Spears of Vancouver found work in the Rosetown district of Saskatchewan, he began to spread the Wobbly message. He was promptly arrested for vagrancy and sentenced to thirty days in jail. His case appeared in the federal Department of Labour's *Report on Labour Organization in Canada*, which expressed alarm at the reappearance of the IWW. It reported that "Spears had been distributing I.W.W. literature and was carrying an I.W.W. card at the time of his arrest."[65]

IWW hopes ran high that they could meet the challenge. In 1923 *Industrial Solidarity* issued a call for organizers to go to Moose Jaw where "Delegates are badly needed as they are very scarce" but where "Sentiment was never better."[66] The following year an organizer in Edmonton promised that "6,000 members of the I.W.W. were coming into the harvest fields of Alberta to show the workers of that province that the eight-hour day could be put into effect in the agrarian

The Industrial Workers of the World offered hope through solidarity.
(*Industrial Worker*, Seattle, 22 July 1922.)

industry."[67] A well-planned campaign dividing the prairies into districts and assigning travelling delegates was presented at the 1925 meeting of the Agricultural Workers' Industrial Union, the first IWW convention held in Canada. Demands were formulated for the next harvest season: "(1) ten hours per day on threshing rigs; (2) minimum wage of $6 per day; (3) transportation to and from the job; (4) blankets to be provided by the employers."[68]

But results failed to live up to expectations. Continuously under the watchful eye of the police, Wobbly organizers in Canada could do little. Saskatchewan Provincial Police found the easiest method of dealing with them was simply to keep them "on the move as much as possible." The Alberta Provincial Police handled them "without gloves." The Alberta police deplored the lack of legislation "dealing with this class of gentry," regretting that "the vagrancy act does not bring them in this scope of the code, and to a certain extent the police are powerless unless we can charge them with mischief or some other misdemeanour." In fact, vagrancy charges were frequently laid against organizers who were "raising trouble" among harvesters. "The only remedy for this condition is through the police," wrote Saskatchewan Premier Charles Dunning to one of his constituents, "and you may have noticed in the press . . . that a number of such men have been arrested by the Regina police for vagrancy."[69]

Arrests continued. The "Canadian harvest attracted quite a number of our members," Wobblies were told at their general convention in 1924, "and arrests over the line were more numerous than last year."[70] In Alberta a number of harvesters, some of whom "had I.W.W. literature in their possession, and who it was asserted were endeavouring to create discontent among the harvesters," were convicted in Alberta of vagrancy. IWW headquarters in Chicago hired lawyers to appeal the cases, and three out of four convictions were quashed.[71] Such successes may have heartened Wobblies, but they were costly in both finances and time, reducing the available resources for organizational work.

Insubstantial though they were, vagrancy charges were a powerful weapon. Even a few days in jail spelled disaster for a hired hand who needed to supplement his yearly farm income with the high wages offered at harvest. Provincial police in Alberta declared with confidence that their ability to secure "a number of convictions" on unspecified charges would have "a salutary effect" on organizers in the harvest fields.[72]

IWW organizing activity among farm workers peaked in mid-decade, then declined. In 1925 a new agricultural unit was formed in Winnipeg but folded the next year. Strong state controls exacted a heavy toll from the Canadian labour movement in the 1920s, and farm workers were only one of the groups to suffer.[73] With leadership in disarray and radicalism suppressed, Wobbly organizers were able to achieve only local and temporary results, seldom more than "agitation, and disruption of work."[74] Farm workers did not achieve effective unified action.

Relations with the Labour Movement

Even more difficult was forging links with the larger labour movement. Workers in resource and construction industries travelled a seasonal route from job to job, from industry to industry. Farm work was a stop along the way. High harvest wages were the obvious draw, but workers might be working to improve their own strategic position, as had the Beverly, Alberta, miners in 1923. "At least 5000 more men than ever before would be needed in the harvest fields," declared Red McDonald of the United Mine Workers of America. "There would never be so good a chance for the miners to enforce the claim for union conditions."[75] While this kind of job itinerancy might present the opportunity for co-operation between farm workers and their counterparts in other industries, it needed an initial link. Without communication and close ties with other labour groups, farm workers could find their strategic position undercut. A similar instance had occurred in the United States the previous year. In 1922, the striking railroad shop craftsmen of Chicago planned to sustain their action by seeking harvest work. Their plans drew an angry response from the Agricultural Workers' Industrial Union of the IWW:

> From the tools and minions of the employing interests we expected no other treatment than what we have received; but from workers, particularly striking workers, we expected an understanding sympathy and active co-operation. It is with pain then, and with some surprise, that we learn of striking railroad shopmen, *through their organizations, preparing to invade the harvest fields with a total disregard of our efforts to win living wages and decent working conditions.*[76]

When it came to the more difficult task of actively establishing a labour organization of farm workers, most unions were either ineffective or indifferent. Neither the traditional trade unions nor, except for the IWW, the fledgling industrial unions directed much attention to labour in agriculture.

Industrial labour organizations were fighting for their lives and had little energy and few resources to extend their efforts to the least rewarding field of labour. The difficulties the IWW faced in its attempt to organize farm workers during the First World War and in the 1920s were considered insurmountable by the other major industrial labour organization in the West, the One Big Union. The OBU's initial attempts to reach farm workers were directed at itinerant harvesters. "With [a] little display of solidarity," promised an advertisement in *The Search-light* in 1920, "we will enforce during the harvest the $6.00 minimum for an 8 hour day."[77] But these forays became part of a struggle within the OBU over leadership and direction of the new union. The lumber workers' unit gathered strength and extended its efforts into lumber mills and harvest fields. Changing its name to the Lumber, Camp, and Agricultural Workers' Department of the One Big Union was a signal to Victor Midgley, executive secretary, that it was encroaching on territory that rightly belonged under the direct control of the OBU.[78]

The lumber workers saw their move as the logical solution to the itinerancy of western workers. Industrial unionization was the only vehicle that could "provide the machinery whereby a proper and efficient organization of the 'migratory' worker can be effected," they argued, asserting that "one unit embracing all migratory workers is the spirit of today." Until the farm workers were "sufficiently organized, or rather numerous enough throughout the Prairie Provinces for the formation of a separate Agricultural Unit," they were to remain "under the auspices" of the expanded lumber workers' unit.[79] The issue was put to the test at the Port Arthur convention in 1920 when the lumber workers asked for the submission of a question to the membership: "Are you in favor of maintaining a Lumber, Camp & Agricultural Workers' Department of the One Big Union?"[80] The Central Executive Committee concurred and the question was carried, but Midgley used the financial delinquency of the lumber workers to have them expelled.[81]

Farm workers were casualties of the fallout. OBU efforts on their behalf appear largely rhetorical. Agricultural labourers were exhorted to organize themselves:

> Now, it is up to the farm-workers and others in the prairie Provinces to line up with us, or rather line up with themselves. Don't wait for a Moses to lead you out of bondage; he can not come. Get together! Call meetings of your fellow-workers in your nearest town or village, and form Branch Units. If there is [*sic*] not enough of you to do so, any worker willing to assist his class can become a delegate (shop steward) and get his fellow-workers signed up in this Union, then get them organized! It does not take education; it takes a man with backbone.[82]

For its part, the OBU seems to have given little more than lip service to organizing farm workers. In 1924, peak year of the decade for IWW activity, the OBU spent more time discussing the problem than coming up with any concrete action. In February, the General Executive Board spent "some time" addressing the matter of "arrangements to handle the harvesters' proposition" for the coming summer. A chain of command was proposed, with instructions to

> again get in touch with Com. Roberts, asking him to forward any suggestions he had as to the best method of tackling this question, and that the Board Member McAllister by [*sic*] also asked to forward any suggestions he might have as to how to handle and organize this class of labor.[83]

In May, the board asked the *OBU Bulletin* to run a weekly notice "calling upon all class conscious workers who were contemplating going into the harvest fields to get in touch with the General Secretary of the O.B.U." It was not very hopeful of results. By way of inducement, it added a cryptic rider that respondents "would hear something to [their] advantage" and planned that in the dubious event that "any answers were received in response to this appeal" they might be considered as job delegates.[84] In August, Tommy Roberts, the leader of the Sandon, British Columbia, branch, the only "healthy and functioning" OBU unit in the West,[85] asked the Joint Executive Board about the organizing activities of the IWW in the Saskatchewan harvest fields. His question "was discussed for some time" but it was not followed up until he goaded them into action with another inquiry in September. The Board then planned to make inquiries of the Moose Jaw delegate, since it was learned that the IWW regularly opened a hall there "for the purpose of capturing the migratory workers."[86] The matter of organizing harvest workers was raised again the following spring, when Board members gave the whole matter a "thorough discussion," concluding that "a plan should be worked out at a very early date."[87]

Organizing in the agricultural industry was clearly a low priority for the OBU, but logistical problems were not the only reason. The OBU had little regard for workers who did not readily join the union. "The task . . . in showing these slaves their class position is almost beyond description," declared the *OBU Bulletin* in 1925.[88] By 1928, the OBU was ready to throw in the towel. After a summer of inaction, the Joint Executive Board decided "that it would be foolish to put an organizer in the fields to organize harvesters at this time." Instead, it continued to rely on members who might engage in harvest work themselves to act as job delegates, receiving "a bonus of 50¢ for every dollar which they procure for initiation fee[s] for new members."[89]

It is difficult to escape the impression that the OBU had little use for workers who did not flock to its fold. Exhortations insulted the very men the union was urging to organize: "Use the grey matter in your skull once in a while and you will realize that by combining you can improve your conditions a great deal. . . . Don't shirk your duty." Calls to action were often couched in language that underscored the gap between urban-oriented leadership and prospective rural constituents. Heroic rhetoric presented a romanticized and unrealistic assessment of the living and working conditions on most farms, and little of practical value for hired hands:

> We appeal to your honor as common laborers. . . . The country still relies on your muscle, sturdy laborers. At Sunset when you ride home on the bare skin of your heavy steeds, caparisoned with the entanglement of the work harness, think of the beautiful picture you make: The Knights of Toil coming back home in the dusk after a day of the great task on which the country relies. You are the builders of an immense wealth.[90]

Ultimately, farm workers were told to organize themselves. "Get your share out of that tremendous wealth," insisted the *Bulletin*, "not only in the shape of wages, but also in proper conditions of work, in good grub and decent bedding accommodation."[91]

These were minimal demands and of the type generally sought by established trade unions. But if farm workers were unable to find a satisfactory place within industrial unions, their position in trade unions was simply non-existent. Traditional trade unions had never expressed interest in unskilled, poorly paid, and transitory agricultural labourers. During the 1920s, prairie farm workers were doubly excluded. As labour began to move into politics, farm workers found it even more difficult to forge links with the larger labour movement. Instead, labour sought an alliance with another dissident prairie group that had begun a similar foray into the political arena – farmers.[92]

The labour movement was recovering from the backlash of the 1919 strike year. In the climate of strident attacks by business and governments on the organized strength it had achieved and the radicalism it had displayed in the previous decade, labour was attempting to consolidate its position. Its own movement into politics was weak, leaving it the option of an ineffective solitary voice or of throwing in its lot with a more powerful group, with the danger of having its own voice drowned out in the process. As Reginald Whitaker explained:

> The most politically advanced elements of the working class movement were forced to choose sides in a struggle predominantly featuring the large bourgeoisie of central Canada, along with its middle-class allies, against the petit bourgeoisie led by the independent commodity producers of the western wheat provinces and rural Ontario. Not surprisingly, they chose the side of those farmers struggling for what was called 'economic democracy,' under the leadership and ideological hegemony of a class whose interests and

outlook were significantly different from those of the organized working class.[93]

The basis for an alliance or even for co-operation between labour and agriculture was by no means substantial. Both groups could agree they had common enemies in large business interests and the railways, which skimmed off economic benefits that were rightfully theirs, and in the federal government, which did nothing to protect them. But on issues closer to home, such as maximum hours of work, workers' compensation, and daylight saving time, they disagreed sharply.

The agrarian movement was supported by farmers who saw themselves as a distinct class, or at least as a separate group with their own economic interests to protect. In deciding to do this through direct political action, they had to make some choices about where to court support. Almost entirely, and in keeping with their vision of co-operative rather than conflictual class relations, they chose not to extend their constituency beyond their own members. They might be willing to co-operate with labour, but they would make no concessions. An editorial in *The Farm and Ranch Review* in 1920 illustrates the suspicion underlying the tenuous relationship:

> Agriculture and Labor can walk a common path just so long as the latter confines its activities to improving the existing order of things by legitimate constitutional means and with due consideration of the interests of other classes. . . . As a matter of humanity, the farmer wants to see all classes able to live in comfort and decency, but that is as far as his real interest in organized labor goes.[94]

When farmers created the Progressive Party in 1920 for their assault on federal politics, they made few active attempts to gain the support of labour. They preferred to concentrate on issues that affected themselves directly and to leave labour to represent its own interests, supporting the Progressives or not, as it saw fit.

In each of the prairie provinces, there were different possibilities for affiliations between the labour and the agrarian movements. Manitoba was still reeling from the General Strike in Winnipeg, which had received open opposition from farmers' organizations. The *Grain Growers' Guide* had denounced the strike for advocating "the doctrines of Bolshevism, confiscation and rule by force."[95] Although there was a good deal of rural support for the aims if not the tactics of the strike,[96] labour was still licking its wounds. It eyed the agrarian movement suspiciously and, as it turned out, with just cause. When the United Farmers of Manitoba entered the political arena in 1921, it severed what tenuous ties had existed with the labour movement. In Dauphin, for example, where a labour candidate had been elected with agrarian support in 1920, the UFM in 1922 repudiated the association.[97]

In Saskatchewan there were more signs of co-operation, although that was an acknowledgement of labour's weakness in an overwhelmingly agrarian province. Some farmers' groups actively supported not merely co-operation but partnership. In 1921 at Ituna, Saskatchewan, the Farmers' Union of Canada was organized, immediately opening lines of communication with the One Big Union and declaring as its slogan: "Farmers and workers of the world, unite."[98] But these attempts at co-operation did not find backing at the local constituency level. Labour candidates either ran under their own banner, in 1921, or, in 1925 and 1929, as Labor-Liberal candidates.[99] Liberal hegemony in Saskatchewan remained unbroken until 1929, with the election of a Co-operative government that relied not at all on a farmer-labour alliance.

Alberta seemed the likeliest spot for concerted co-operation between labour and agriculture. By late summer in 1919 the United Farmers of Alberta was committed to enter politics as a distinct economic interest group. It sought co-operation with labour in order to fight a common foe, but its ally was fully expected to look out for its own interests. The newly politicized agrarian movement could benefit from its ties with labour, but it did not need to make concessions to do so. Labour, on the other hand, lagging far behind the farmers in numerical and economic strength,[100] made overtures to the agrarian movement that had a decidedly conciliatory slant.

The *Alberta Labour News*, which billed itself as the "official paper of organized labour in Alberta," and which was the unofficial organ of the Alberta branch of the Canadian Labour Party, warmly greeted the entry of the United Farmers of Alberta into the political arena. In its inaugural issue it sought the support of farmers for its own aims and soon promised the co-operation and support of the Alberta Federation of Labour for the UFA in the upcoming election.[101] During the 1921 campaign, labour candidates identified themselves closely with the UFA, promising that "the next Government will be a Farmer-Labor Government."[102] Throughout the decade the *News* pressed for farmer-labour co-operation, seeking it particularly at election time.[103] The results of that decision redounded upon farm workers.

The strong stand that the *News* took on issues affecting labour was watered down when it came to agricultural labour. Farm workers were seldom mentioned in the pages of the newspaper. When they were, the thrust of the comment was not concerned with wages or conditions of farm work but rather with reconciling the interests of farmers and labour. Immigration was a case in point. Both farmers and workers agreed that immigration of the unskilled kept wages down. To farmers this was a benefit, but to labour it was a disadvantage. The *News* tried to resolve the difference by pointing out how few of the recent immigrants were going to farms. In 1927, it showed that only fourteen out of 215 newly arrived men were reported to have been directed to farm jobs from the Edmonton Employment Bureau, and of the 1,540 orders for spring farm help, central Europeans filled only six. In Lethbridge, none of the 139 immigrants who received work were sent to farms. The situation was similar in Medicine Hat and Drumheller.[104] Moreover, the *News* assured farmers that the recent immigrants were unsuitable for farm work. Central Europeans especially were unfamiliar with prairie agriculture and were unable to speak English. "Farmers applying for help stipulate that they do not want foreigners," declared the *News*, and "farmers simply will not have men who cannot understand simple commands."[105]

In the meantime, the farmers used their organizational network to attempt to regulate the farm labour supply and conditions of employment and to curb wages. The United Farmers of Alberta struck a farm labour committee in 1920, when wages were still buoyed by post-war demands. "Farmers are unanimous at the present time in saying that we have been paying entirely too much for farm help," declared W.D. Trego, committee convener, urging the members to use their organization to exert its influence. "Up to the present time the U.F.A. as an organization has taken no part in regulating farm wages," he said, "but if our members are to continue to produce, it is evident that we must come to a decision as to what we can pay for labor."[106] In early 1921, it concluded that $40 per month ought to be the base wage, well below that of the previous year when Alberta farm hands had received an average of $58 per month for year-long work.[107] An inducement was suggested in the form of a $5 or $10 bonus "in case the farmer secured sufficient crop and sufficient prices to make him 6 per cent. interest on his investment in land and equipment."[108] Farmers across the prairies united to tackle the issue. In March, 1921, they held a conference of western farm organizations in Regina and spent "the whole of one day" discussing wages. They agreed to a prairie-wide rate of $60 per month for "fully experienced men if they remained throughout the season," the rate to be dropped by $10 per month for men who "quit of their own accord before the end of the season."[109] Throughout the decade, farm organizations across the prairies worked to keep the supply of labour high and wages low. And in its search for support from the larger and better organized farmers' movement, labour voiced no objection.

On rare occasions, farm workers were included in the general protests launched by organized labour and thus received its backhanded support. The Alberta Federation of Labour had elicited a promise from the provincial government to end the practice of leasing out prisoners and denounced its failure to do so. At its 1924 convention, the Federation resolved that it "strenuously protest to the Provincial Government against the continuance of leasing out prisoners to work for farmers and other employers."[110] Similarly, the Federation resolved to urge the federal government to amend vagrancy sections of the Criminal Code. It objected that workers should be found liable for conviction "simply for turning down jobs with low wages."[111] The lowest-paying jobs were nearly always farm work, and police pressured unemployed men to take them. In its 1930 reiteration of the resolution, the AFL bolstered its demand by citing the example of several men who had been sentenced to the Fort Saskatchewan penitentiary "on the charge of vagrancy when the evidence showed that they had only been in Edmonton two days, but had turned down farm jobs offered them at low wages."[112] Yet the incident had occurred two years previously. At the time it had passed unremarked, and it did not elicit any suggestion of protection for farm labour wages.

In 1928, the International Hod Carriers, Building and Common Laborers' Union No. 92 of Edmonton, a Communist Party-led union,[113] succeeded in passing a resolution through the Alberta Federation of Labour "favoring the organization of the Agricultural Workers as a part of our movement" and asking the Trades and Labour Congress "to take steps to bring about" such an organization. The purpose, according to the hod carriers, was to ensure that the "trade and occupation" of farm workers "may be elevated to a higher standing and that they may receive the benefits of legislation now already enjoyed by other workers." These were laudable aims, but the resolution was based on the concern that changes in agricultural technology had made the industry "more dependent upon skilled and efficient labor," while the federal government and railway and colonization companies were "yearly dumping thousands of immigrants into this country under the disguise of farm laborers."[114]

The effect of restricting immigration to agricultural settlers and farm workers, combined with the Railway Agreement, was to create a labour surplus. By the end of the decade, it was turning into a glut. Labour's concern about the plight of the farm worker is suspect. The blame for poor conditions of labour in agriculture was laid squarely at the feet of the "the Farmers' Organizations and Co-operative Societies [in which] no real consideration has been given for the economic betterment of the so-called 'farm hand' or agricultural worker." The AFL shunted onto farmers the responsibility for the exclusion of farm workers from labour legislation such as compensation.[115]

Farmers were merely protecting their own interests. In the case of workmen's compensation, farmers protested vigorously against the inclusion of their own employees. Henry Wise Wood voiced a number of objections while the subject was being investigated in 1917. "I think there would be a great difficulty in the practical working out of it," he explained, "on account of the nature of the relationship between the farmer and the labourer." The variable length of employment, ranging from a day to a year, the lack of control that a farmer had over the price he received for his product, and his inability to add the insurance costs "or get it from the consumer" all made the scheme unworkable.[116] Labour appears to have concurred. Despite an International Labour Organization draft resolution in 1921, Workmen's Compensation Acts in all three prairie provinces specifically excluded farm labour, even when the statutes were revised after the resolution.[117]

The fragile unity of interests between the agrarian and labour movements was achieved at the cost of the men who straddled both groups – farm workers. As Communist Party organizer J.M. Clarke explained:

> Theoretically a number of farmers favor co-operation with labor organizations, and many are inclined to view organized labor as an ally of the working farmers in their struggles against capitalism. But to suggest that this unity of action should be extended to an alliance with the hired man –. Well, I am afraid it would not be received with open arms except in a few isolated cases. Let us never forget that practically all Western wheat growers are exploiters of labor at some season of the year, and . . . this tends to foster a class ideology. It is one thing to speak of co-operation with organized labor in the cities, where the class demands of the workers do not directly affect the farmer's position; it is a different matter to speak of co-operating with the man he is exploiting. The formation of an alliance between poor farmers and proletarians is not quite so simple, nor so mechanical in Western Canada.[118]

Certainly, farmers did work to protect their interests with regard to their own hired labour. They could not be expected to support measures that would cost them money or labour-time. While the small farmer might be sympathetic to the labourer in construction or bush work or mining, especially if he had to supplement his farm income or to underwrite his farming venture by waged work in these industries himself, he drew the line at extending provisions for health and safety protection or minimum wages or paid holidays to his own employees. Farmers who were managing their budgets close to the bone could ill afford the costs of workers' compensation.

The idea of maximum hours on farms drew the greatest criticism from farmers, who scoffed at the idea of an eight-hour day, pointing out that farm work did not move to the rhythm of the factory. Animals had to be tended early in the morning and late in the day, and at harvest-time the demands of ripening crops and the threat of early frosts dictated working days of twelve, fourteen, or even sixteen hours. Farm workers who inquired about hours of work were curtly told that such matters were subject to "the custom of the district" or, in the absence of any established custom, that they were simply to obey their employer.[119]

When Saskatchewan Premier W.M. Martin was invited by federal Minister of Labour Gideon Robertson to send a delegate to the International Labour Conference in 1921, he declined on the grounds that "it is very doubtful as to whether any good can be accomplished by endeavouring to have the Conference agree on the question of the regulation of the hours of work in agriculture." He pointed out the great variations in types of agriculture, systems of cultivation, and climate, questioning "how any such legislation is possible." His conclusion was that, especially at seed-time and harvest, "a short hour day seems to me to be simply out of the question."[120] When the International Labour Organization went directly to the United Farmers of Manitoba for a response, it was informed in no uncertain terms that "the great majority of farmers in Canada appear to be of the opinion that an eight hour day on the farm would make profitable agriculture impossible, and if such proposals were seriously considered I am sure that you would hear from us."[121]

Nor were farmers particularly sympathetic to calls for better working or living conditions for their workers, since they had once endured the same conditions themselves. Farmers reasoned that there was no logical basis for providing their hired hands with better conditions than the ones with which they themselves had had to contend. No other industry provided its work force with even equal, let alone better conditions than those of its employers. Some farmers carried the argument even further, declaring that the hired hand owed a debt of gratitude to his employer that transcended the wage agreement. One farmer's wife was explicit:

> In a new country like this, where every man – the farmer as well as his 'hand' – is himself a labourer, to hold the hard worked and harassed farmer at harvest time for wages that he cannot afford to pay and continue to function save at a loss, is a poor return for a sincere effort on the farmer's part to give a home and a living to the stranger within his gates.[122]

And in those cases where farmers' conditions were superior to those of their employees, this was considered only fair. After all, most of them had started out in the same condition, and they had put up with hard work and pioneer hardships. There was a pride and even status attached to those who had suffered and survived extreme hardship. Moreover, the rewards of that hardship were directly translated into material improvement for the farmer. As long as it was believed that hired hands could win the ultimate prairie prize of farms of their own, their own hard work and thrift could be expected to translate into their own material improvement.

But such a translation could not be easily achieved in the post-war period. Farm workers in the prairie West were now a proletariat. Yet, when they contemplated proletarian paths to material improvement, they found that governments and private agencies were formidable foes and that unions were ineffective or indifferent allies. Hired hands found that their most practical strategy was still to rely on their own resources.

8

The Dialectic of Consent and Resistance

"You're working for yourself you might say."[1]

Jens Skinberg was drawn to the Canadian West from his native Denmark by the anticipated independence of farm ownership, but by the time he arrived in the late 1920s agricultural labour no longer provided an easy entry. He continued working as a hired hand. The autonomy of his job gave him a good deal of satisfaction, creating the illusion of "working for yourself." Yet Skinberg was not simply indulging in delusion. His conviction was tempered by realism about the labour involved and resonated with his own ability to influence the nature of his work and social conditions.

Skinberg had come to the prairie West with every intention of becoming a farm owner. He expected his stint as a hired hand to be a short one while he accumulated the necessary capital to start up on his own. But the closest he came to operating his own farm was the short period when he rented a half-section in partnership with another man. For many similarly disappointed newcomers, the alternative was to leave agriculture altogether. In fact, in his years as a farm labourer, Skinberg at times drifted off to Vancouver and "various other places" in search of winter work. But the spring planting season invariably drew him back to the prairies. He "moved around quite a bit" in his pursuit of farm work but never seriously considered abandoning it.

Despite the hard work and low pay, Skinberg chose agricultural labour for advantages he did not find in other lines of work. He found a ready welcome in the agricultural community that was not duplicated anywhere else. He returned year after year to work on the same farms and was part of a social network that encompassed all elements of the rural society. As a member of a social club in his adopted community of Dalum, Alberta, he was invited to "a fancy dinner" as often as once a week. Although membership in the club was supposedly restricted to men who were over thirty years of age and owners of property, Skinberg was a member in good standing and enjoyed the "drinking, poker, dancing or hymns." He was proud to be part of a band that would play free of charge for "various little do's."

The close network of rural society was reinforced for Skinberg by kinship in his ethnic community. The conviviality of "Danish traditional entertainment" was an important feature of the social life. Each year the club would hold a party for all the local families and regale them with spontaneous songs about local events and people, a favourite Danish pastime. Skinberg's working life was also centred in the Danish community. He found most of his farm jobs within the district, and even when he worked farther away in the sugar-beet fields near Raymond, Alberta, he returned to Dalum as soon as he heard the crops were ripening. Skinberg's loyalty was rewarded by mutually satisfying relations with his employers.

The work, too, held its own rewards. As he learned to master the new machinery and equipment, Skinberg derived satisfaction from his evolving skills and found a high demand for his proficiency and experience. His independence was nourished when he enjoyed the latitude to "sit down if I felt like it." He learned how to negotiate his own pace and schedule, choosing whether or not to "work after supper."[2] Skinberg spent his years in the prairie West as a hired hand out of choice, finding in the farming community and in the work sufficient satisfaction to tie him to agricultural labour. His experience illustrates that of the men who laboured in agriculture at a time when their position was becoming more fixed and their relations with capital were taking on new dimensions.

Farm Workers and Rural Society

The relationship between labour and capital became more conflictual during the 1920s, yet the conflicts were played out in muted form. As hired hands negotiated the changes of the decade, they arrived at a delicate balance between consent to their new position and resistance to their conditions. In moving toward consent, they adjusted by internalizing social constraints. The social side of labour-capital relations in prairie agriculture thus takes on particular significance. The relations were forged at the point of production but extended well beyond it. Agrarian culture and ideology were powerful constraints in mitigating the conflictual relationship.

Productive relations determine the basic contours of social relations. However, the nuances in and between each relationship are coloured by the configurations of other affiliations.[3] Although hired hands were not members of the farm family, they were accorded a special position within the farm household and within the farm community. To a greater extent than in most other industries, farm workers developed a relationship with their employers and with agriculture as a whole that, although firmly rooted in the relations of production, stretched far beyond the purely economic.

With logistical barriers to forging links with the working class and with economic reasons for finding common ground with the dominant class, agricultural labourers saw themselves as agriculturalists first and as labourers a distant second. In pioneer communities, the sparseness of population led to necessary co-operation, while the availability of land led to a common sense of purpose. Hired hands found a welcome in the dominant culture. This alliance appeared to continue during the 1920s, and the social relations served to mediate their basically conflictual relationship with capital.

Yet the relationship itself was in flux. As agriculture moved from a pioneer economy toward a greater involvement in a market economy it became more enmeshed in the structures and style of capitalist production. The culture was thus becoming a complex amalgam of traditionally rural and agrarian values and institutions and those emerging from the pioneer era that sought to embrace the objectives of maturing capitalism. Farmers hoped to achieve the best of both worlds. Wheat pools were but one attempt to "create a new form of enterprise which blended economic viability with social purpose."[4] Farm workers adopted tactics to bring the best return for their labour as they sold it on the open market, yet looked, too, for employers with whom they could get along. A difficult farmer would soon find that his "field of labour is narrowed down to strangers and those who have not yet heard of his reputation."[5] The result was a combination of often idealized notions about social relationships with hard-nosed economic strategies on the part of both labour and capital.[6]

In the social realm, this combination produced diverging ideologies. The view of bachelorhood is a case in point. As noted in Chapter 5, bachelors, whether homesteaders or hired hands, found a ready acceptance in the family-oriented community of the prairie West. This example of farm workers' acceptance of the dominant culture in exchange for equality of social relations helped to balance the economic disadvantages of the apprenticeship system during the pioneering era.

The bachelor identity had been constructed to suit the needs and to fit the ideology of a pioneer agrarian community. But as that community moved toward capitalist values, with increasingly divergent aims between labour and capital, the identity was reconstructed to suit the economic aims of both the individuals who carried it and those who observed it. Of particular importance for bachelors during the original construction of the identity was that even though the ideology was rooted in material conditions, an individual's actual economic position did not figure highly in whether or not he was accepted into the society. It likewise did not figure highly in the identity of bachelors. Most were not wealthy, but their capital-poor position was to be a temporary phenomenon. Thus the link between the economy and the ideology, although very strong, was blurred during the pioneer era and, consequently, in the identity that bachelor homesteaders constructed for themselves.

But the pioneer era did not last. Homestead lands filled up; purchased lands were taken; more land came under cultivation; and economic and social institutions were established and matured. Sparsely settled and underdeveloped pioneer areas made the transition to mature agricultural communities. A significant marker of the transition was the increase in the number and proportion of farm families. By 1921, more than 55 per cent of the rural male population aged fifteen years and older was married, bringing the prairie marriage rate into line with the Canadian average.[7] The ideal of a family-oriented economy and society had become the reality.

With the growth of capitalist agriculture and the transition to a fully developed commercial economy, social stratification began to reflect the growing economic stratification. Social distinction based on economic position was a phenomenon that had been denied and deplored during the pioneering years and had been held in check by the sharing of pioneer hardships. But it came to be recognized and even – somewhat sadly and reluctantly – accepted as the price of progress.

At the same time, entry into farming was becoming extremely costly, and thus extremely difficult. This did not stop the movement of single men onto the prairies, but it did change their position. Instead of being bachelor homesteaders, they were now bachelor hired hands. The distinction was not always clear-cut. In the early days, nearly all homesteaders and probably most men who purchased farms had worked out as agricultural labourers, but they had been regarded primarily as farm owners. Now there were many more men who were identified only as farm labourers, and they faced seriously diminished opportunities to become farm owners and heads of farm families. As late as 1931, only 15 per cent of waged farm workers appear in the census as heads of families.[8]

The family orientation of prairie agriculture achieves particular importance with the new position of bachelors. As indicated in Chapter 5, the bachelor identity encompassed attributes that were characteristic of masculinity, of agriculture, of pioneering, and of frontiers, but these attributes were not particularly restrained by class position.

However, the cultural construct of bachelorhood came to have a very different meaning as it became defined within narrower economic parameters.

The resulting shift in attitudes toward bachelorhood presents an interesting glimpse of a culture in flux. Aspects of the prairie culture that had provided a special place for all bachelors within the agricultural community came to be applied primarily to the work force. Bachelorhood was now more restricted. It came to be seen as part of the definition of agricultural labourers, in contrast to its earlier application as part of the definition of a great many newcomers to the agricultural community, whether owners or workers.

The cultural change reflected an important structural condition. Bachelorhood was, and continued to be, a precondition of most agricultural employment. Low wages, job insecurity, a high degree of mobility, and a lack of accommodation for families all continued to preclude marriage and family from the lives of most farm workers. Despite various settlement schemes of the 1920s to bring British men with families to the prairies, where they were expected, and in some cases required, to undertake farm labour as a condition of their assistance, the numbers of married farm workers remained low. In 1931, only two out of fifteen farm workers and farmers' sons were married, compared with almost three out of four prairie farmers.[9] According to the federal Department of Immigration and Colonization, "nine out of ten farmers want to have nothing at all to do with a [farm hand's] family." Although some farmers welcomed the services that a farm hand's wife could provide, most saw a labourer's family as trouble and expense. "It is only the one out of ten that will even consider giving employment to a man and his wife, and the number who will take care of man, wife, and several children is much more limited still."[10] One farmer went so far as to refuse to honour his agreement to take on a hired man when he discovered that the man's wife was pregnant.[11] It was a circular process that was reinforced as bachelorhood came to be associated with hired hands. Farmers often recognized the advantage of employing married workers whom they regarded as steadier and more reliable, but by far the majority of them hired single men.[12]

Judgements about bachelorhood began to change.[13] In the early years, bachelor farmers were expected eventually to settle down, own a farm, marry, and raise a family. Those who had not yet done so were excused on the grounds that they were still economically unprepared to support a family or there was a shortage of women. But as the male-to-female imbalance was redressed and as farmers began to make economic progress, these reasons lost their validity. And as bachelorhood came to be more restricted to hired hands, it was perceived as evidence of an unambitious nature. Manitoba's Department of Agriculture began to characterize farm workers as men "who have failed in pretty nearly every walk of life."[14] Advertisements in farm journals contrasted the "ambitious man" with those who were satisfied to pitch hay for $4 a day or do chores for $3.50.[15]

Hired hands were often seen as too shiftless to settle down and own farms or raise families. Those who were not making definite plans to take up farming on their own account were dismissed as "drifters."[16] And since they had little opportunity to become farm owners themselves, they were no longer eligible to marry the farmer's daughter and thus solve their dilemma. A contemporary observer explained a common perception that placed the farm hand in a double bind:

> As the labourer nears thirty-five years of age, at which period in his life he should have accumulated substantial savings, he approaches a transitional period. This phase is noticeable for his desire to be independent. If the labourer fails to show this characteristic restlessness and settles down to the acceptance of agricultural employment as a life task, he has past [sic] beyond that period of most effective service and has entered the final stage indicative of the stagnatory period.[17]

And farm workers were getting older. By 1921, more than half were aged twenty-five and over. By 1931 the proportion had increased even further.[18]

The status of bachelors declined. The earlier bachelor homesteaders had been accepted into the dominant culture because as they constructed their identity, they both appropriated and embodied features of the prairie ideology. It was in their interest to do so, since social acceptance was an important component of agricultural success. As well, they embraced the ideology because it coincided with the type of future they sought for themselves. They shared the majority vision of agricultural progress based on sacrifice, hard work, and a willingness to endure harsh conditions.

But bachelor hired hands did not seem to adopt this view. Their interests, and those of the dominant ideology, began to diverge. They were not eager to bear the risks of agriculture for the dubious possibilities of future success. Hired hands would have scoffed at the proposal of the disgruntled Saskatchewan farmer who wanted "some system for dividing with the hired man under which the farmer's interest will be protected in case of a deficit."[19] They were unwilling to work for the meagre rewards that most farmers offered and to tolerate the wretched conditions. Farmers were urged to recognize that conditions of labour had changed over the past fifteen to twenty-five years. "Labor looks upon life in an entirely different way now than it did then," insisted the *Nor'-West Farmer* in 1920. "Laboring people . . . insist on more pay, a shorter day and more comfort in life than were customary then. Labor is through being driven to work. In the future it must be led to work."[20] Hired hands were working for their own individual betterment and were not willing to sacrifice their own interests for that of the farms where they worked, or for the farmers who employed them, or, it seemed, for the agricultural community of the prairie West.

The economic contribution of the bachelor hired hand to the agricultural community was measured by different criteria than those used for the bachelor homesteader. Even though a bachelor homesteader made slower progress than did a married one, his work on his own land and his potential for developing his own farm were the benchmarks of his success. But the economic contribution of a bachelor hired hand was measured only in his present ability to help his employer bring in a good crop or otherwise improve his employer's farm. His contribution was indirect, undervalued, and suspect. When farmer H.A. Kuhn reported to the CPR that the farm he had purchased from them was not doing well, he was blamed for placing too much reliance on his hired man. "Our experience is that where there are such valuable improvements and equipment as you apparently have on the farm, that it requires somebody who is personally interested in the place, to get the best results out of it."[21]

The economic position of a bachelor hired hand likewise apparently altered his social contribution. Again, it was a shift in emphasis from potential to present reality. As a bachelor homesteader he would have been judged on his probable ability to find a wife who would provide a higher level of culture, sophistication, and gentleness not only to himself but to the community as well. As a bachelor hired hand, however, he was less likely than ever to marry and was thus less susceptible to the expected civilizing influence of a wife and family. His social contribution was often measured only in terms of whether he brought harmony or discord into his employer's home.

The behaviour of bachelors was likewise examined, and received harsh judgement. Women played an important role in this assessment. They were charged with bringing moral and cultural uplift to a male-dominated society and with providing the guardianship and discipline necessary for a morally and socially upright future generation.[22] Their task was to create a climate of social well-being unlike the rough masculine behaviour that had dominated the frontier.[23] While farm women continued to express special concern for young single men who were incapable of taking care of themselves, they also worried about the moral standards of their families. Social conduct such as drinking, gambling, fighting, and visiting houses of ill-repute was much less easily forgiven.[24] This type of behaviour appeared to decline with the advent of families, so its occurrence stood in stark contrast to the behavioural norms of the developing agrarian community.

In the home, personal habits of farm workers were subject to scrutiny. Mothers were instructed to protect simultaneously the morals and the health of their families. "Is the average home sanitary?" asked the women's section of the *Grain Growers' Guide*, deploring the use of the common water dipper. Even trusted neighbours might pass on disease through such a practice, and "Then there are the hired help, about whom you may know nothing. Is it right that they should drink out of the same vessel as you and your children?"[25] Farm women were becoming increasingly concerned to "observe the little niceties and decencies on the farm where we are making a real home and trying to bring up the children."[26] Farm hands were rated on cleanliness, manners, and as an example for the children. Many a bachelor hired hand failed the test. As "A Mother of Two" cautioned:

> I think if a woman does her duty in the home, in the training of her children, she cannot be too particular as to those she admits to the privacy of her home, especially on the farm, where the hired help . . . are bound to associate with and influence the children to a certain extent.[27]

These changes were taking place within the context of a transition in the position of agrarian life in the larger Canadian economy and society. As industrialization challenged the economic and political hegemony of agriculture, as scientific thought and method came to dominate the business world, and as the bright lights and economic opportunities of cities began to lure young farm people, rural life came under attack. The most visible effect was the depletion of rural population in central Canada, a problem that governments, churches, and concerned citizens addressed.[28] But westerners worried, too. Farm journals expressed the common fear that people did not recognize agriculture as the "anchor" of the nation, the "great untroubled reservoir of sanity and common sense."[29] Farm people grew alarmed that agriculture could not hold on to young people. The finest, who possessed "as much courage and capacity for hard work as their forbears," were finding that "the game has not been worth the candle" and were being drawn to the cities by "the spirit of adventure, which is the will to success."[30]

The agrarian community responded by formulating and advocating a "Country Life Ideology" that counterposed the land as "the fount of health, peace, virtue and the Spirit" to the city as "unproductive and parasitic, [the] source of dissipation, sham and unfulfillment."[31] This ideology took hold during the first two decades of the twentieth century and shaped the attitudes and actions of farm dwellers throughout the

1920s and beyond. It combined with a growing emphasis on the role of women as homemakers. Women were urged to recognize that their responsibilities toward their families included emotional, psychological, and physical well-being. A farm woman from rural Saskatchewan acknowledged the task. "We of the farm must recognize the fact that there must be time and strength for social, religious and intellectual interests."[32] The farm home was expected to be a "haven of safety and healthfulness."[33] A contest run in 1922 by the women's section of the *Grain Growers' Guide* asked farm women: "Do you want your daughter to marry a farmer?"[34] Their answers expressed fulfilment in a life close to nature and free of the social hazards of the city environment. They also expressed desire for an intimate family life that by definition included only members of the family circle.

Bachelors came under fire for failing to live up to the agrarian ideal. Their very bachelorhood served to cut them off from membership in the institution extolled as the heart of country life. The nurturing and reformist aspects of the ideology identified the family farm as the source and haven "of lasting democratic values, and the structural basis of . . . egalitarian rural communities."[35] The agrarian defence of these values elevated the importance of the family, which was seen to embody the rural virtues. In such an ideological climate, there was only a restricted place for bachelor hired hands.

They were very visibly just what city-folk were saying was wrong with the country. Their exclusion from family life made them country bumpkins and their low wages made them poor country cousins. In one study, the farm worker of ten years' duration was characterized as "lack[ing] sufficient education or initiative to rise above his present situation or . . . mentally incapable of doing so."[36] Moreover, their vision seemed to be narrowly focused on attaining higher wages, not on becoming more efficient producers. Farmers complained that they must have workers of "a class which will work to the interests of the farmer."[37] Hired hands were not recognized as making a significant economic contribution to their industry, let alone to the Canadian economy.

And given the importance of farm ownership in the prairie ideology, bachelor hired hands were often seen as economic failures, in sharp contrast to earlier bachelor homesteaders who had been viewed as potential successes. This was an important distinction in an industry that offered no guarantees. In the basic insecurity of prairie agriculture, with the uncertainty of rainfall, frost, and disease, the threat of crop failures and foreclosures, and in the context of a soaring cost-price spiral, farmers in next-year country were haunted by the spectre of sinking into agricultural waged labour.

The bachelor identity came under attack. There were two dimensions to this reassessment and harsher judgement. First, there were real changes in what was perceived by both bachelors and outsiders as constituent elements of the bachelor persona. The economic changes of the post-pioneering period brought about most of these changes and signalled a deterioration in the social and economic position of bachelors. Second, tolerance diminished for what was perceived as typical bachelor behaviour. What was formerly seen as healthy independence was now seen as irresponsibility. What had been an endearing quality of inability to care for oneself was now slovenliness. What had once been tolerated as youthful letting off steam was now imprudence. Hired hands were subject to stern judgements: "From what I can learn, any work that Traynor may have lost has been his own fault," declared the CPR's Assistant Superintendent of Colonization about a hired hand. "He comes into town whenever he gets an opportunity, and spends most of his time in one of the beer parlors, gets full up and then becomes nasty."[38]

In response to these changes, bachelors began to reconstruct their identity. They focused on different attributes, giving new emphasis to some while discounting others, or they reordered them in consideration of their changing importance, or they used them in different ways. With the opportunities for farm ownership so curtailed, bachelor hired hands focused their energies on immediate issues. "From the farm labourer's point of view," according to the federal Department of Labour, "he is interested first in a suitable wage, next, in reasonable hours for labour and then in stable employment."[39] Unlike bachelor homesteaders who were often as much, or even more, interested in learning to farm as in accumulating capital, bachelor hired hands concentrated on improving their wages and their working and living conditions and on adjusting to their new economic position. In doing so, they called on old attributes of the bachelor identity, which had been shaped by an ideology with which they were now at odds, and they used these attributes in different ways.

The construction and reconstruction of identity is a complex process, a response to the practical needs of the group and to the values of the larger ideology. The anticipation of marriage, which had played such a significant role in shaping the identity of the bachelor homesteader, was less a consideration for the bachelor hired hand.[40] Indeed, remaining a bachelor was a positive asset, since the work paid meagre wages, required great mobility, and seldom made any provision for the accommodation of a wife and family. A 1922 United Farm Women of Manitoba survey revealed that only thirty-seven out of the 366 respondents reported a separate house for the hired help.[41] The hired hand who was unencumbered stood a better chance of making a living.

The steadiness that was an important attribute of the homesteader bachelors, the willingness to put off present pleasures for future rewards, was less important, indeed often irrelevant, to bachelor hired hands. When the United States Industrial Commission studied hired hands in 1899, it reported that the breakdown of the agricultural ladder undermined the ambitions of bachelor farm workers. With farm ownership beyond their reach, "there seems to be a decided tendency for the farm laborer, if he is unmarried, to work for money without a very definite object." Any money earned "is likely to be used for what his fancy dictates, most likely for a horse and buggy of his own."[42] A quarter of a century later, Canadian prairie farm wife Kathleen Strange reported that all hired hands yearned for a car. "Automobile agents reaped a fine harvest out of hired men by selling them second-hand cars," she recalled, "and often took the best part of their wages every month in payment for them."[43]

The most significant changes occurred when the men called on old attributes of the bachelor identity and used them in different ways. Independence is a case in point. As indicated earlier, independence was an important attribute of agricultural pioneers, and it held a revered place in the prairie ideology. It is not surprising, then, to see it surface in the identity of bachelor homesteaders, who needed it in liberal doses if they were to carve a farm out of the virgin prairie. And when bachelor homesteaders became bachelor hired hands, the attribute retained its significance, although the reasons for its importance changed and so did its uses.

Farm workers applied their own meaning to the attribute of independence, using it in their own interests and in a way that was inimical to those of their employers. This appropriation of a characteristic of the dominant ideology, and its subversion, was most clearly expressed in the "independence" of hired hands in their economic position. As indicated in Chapter 6, they were notorious for job-jumping, for quitting work at the slightest provocation or for no reason an employer could understand. The independence that job-jumping expressed was an important attribute of bachelorhood and something that farmers deplored – they agreed that married men were much more reliable than bachelors. From the farmer's point of view, single men had a "tendency . . . to wander from place to place."[44] But to the men who were using their independence in this way, it was an effective strategy to discipline bosses, to achieve better working or living conditions, and, most importantly, to maximize earnings.

The seasonal nature of agriculture resulted in jobs that were short term and a work force that was highly mobile. Both factors discouraged collective action and encouraged individualistic responses to wages and working conditions. In the positive sense, farm workers were able to parlay the immediacy of seasonal demands for their labour into high wages. When hired hand Sydney Metcalfe left the farm of Robert Stuart, the CPR farm placement office commiserated. Farm workers "are always looking out for a higher wage," declared James Colley. "Many of them prefer to work for higher wages for short periods and take the chance of being out of work during the winter."[45] Acting individually, farm workers were able to demand and receive high wages when farmers were hard-pressed to bring in their crop and could not afford the time to search for cheaper labour. Farmers were forced to recognize that their hired men might want to "quit forthwith without cause or grievance" when the prospect of higher wages beckoned. "Some men get that way about this season when help is scarce and prospects for higher wages in harvest are good," mused the editor of the *Nor'-West Farmer*, "or [when] somebody offers them a better wage than they are getting."[46] In the negative sense, collective action was simply too risky. Hired hands could ill afford the possibility of a prolonged strike during the brief high-paying season. Especially at harvest, the demands for their labour were so urgent and so strong that an individual refusal to work was often a better strategy than was an attempt at collective action, which took time and organization and was likely to bring down the force of law. It was much faster and more cost-effective simply to scout out better-paying jobs.

Moreover, the nature of farm employment and the organization of the industry encouraged individualistic action. Most hired hands spent far more time with their employers than with other farm labourers, and their work required individual initiative rather than co-operative effort. The many small employers in this extremely time-pressured business could be more quickly and effectively disciplined by immediate direct action, such as job-jumping, than by the slower method of large strikes, which require planning, central organization, money, and time. Hired hands may have had little opportunity to develop the habit of collective action, but their continued use of strategies of individualism was based on past success, both in gaining employment and in exercising some measure of control over the conditions of that employment.

Independence, expressed as job-jumping or in other individualistic methods of dealing with the working-class realities of farm labour, was thus a significant form of worker protest and was sharply at odds with the agrarian ideology in which it had been nurtured. As the interests of the bachelors and the larger agrarian community began to diverge, the men redefined their identity. They carved a new niche for themselves, still within the parameters of the dominant agrarian ideology but more suitable to their own changing needs.

The changes that took place in the 1920s in the social structure of the agrarian community were reflected not only in the institutional forms of the dominant culture but also in ideological developments. This allowed farm workers to maintain a very ambiguous position with a foot in each camp, one in labour and one in capital. They subscribed to the dominant culture in its institutional forms, yet embraced and manifested an ideology that was appropriate to their class position. They were able to do this because they shared with the dominant culture an ethos of individuality, which they appropriated ideologically in an oppositional manner. "Ideology" has a particular definition in the prairie agrarian context. It is used in conjunction with another concept, "ethos," from which it is distinguished in the following way: both embody ideas about standards of behaviour, attitudes, morals, and ideals, but these concepts exist in a two-tier relationship. Ethos is a more generalized concept, less clearly defined, and encompassing a very broad range of attitudes and ideals. Examples are values that can even be contradictory, such as those of co-operation or individuality, both of which were embraced at different times and under different circumstances by prairie residents. These values are not the exclusive property of any particular group. Ideology is more specific in that it selects certain features of the ethos and appropriates them for its own use, or interprets them in a way that is consistent with its own needs. An example of this is the way such values as co-operation and individuality are perceived in different cultures.

In the early pioneering period, and even later, the ethos of individualism was an important part of the value system of the agricultural community. Although coming to the West was often a family or group strategy, individualism was a characteristic that was widely celebrated and to varying degrees necessary for agricultural success. It was espoused by those pioneers who were individualists in the first place, having headed west in the exciting days of hard work and free land and the building of a new society. All members of the agricultural community, whatever their economic situation, embraced the ethos to some degree.

It was also very practically applied in the workplace, where individual initiative, judgement, and resourcefulness were highly valued, indeed essential, components of agricultural success, whether for owners or for workers. The farmer was lauded for possessing these qualities, and he hired men who exhibited these qualities as well. A hired hand who demonstrated initiative, good judgement, and resourcefulness, and who could be trusted to perform a wide variety of farm tasks with little supervision, was in great demand. Thus not only was the farm worker likely an individualist to begin with, since his purpose in coming to the West was to establish a farm of his own, but this attribute was among the most highly valued as he sought employment and the most highly encouraged as he continued in it.

Individualism was certainly used to serve capital's needs, but it also could serve labour's needs. And farm workers used it in ways that served their needs and were antithetical to those of capital. If men were dissatisfied with their employer, their working conditions, or the food on the table, they simply quit. They appropriated the ethos and embodied a particular use of it in their ideology in a way that was consistent with their class position. As a tactic in their struggle with capital, it allowed workers to maintain control over "one's labour power and its disposition."[47] It disciplined employers, it maximized wages, it influenced working conditions, and, on a broader scale, it even affected the rate and direction of technological change. It was a vital, aggressive form of worker protest, and one to which farmers were forced to respond.

Farm Workers and the Labour Process

From the point of view of farmers, agricultural labour was one big headache. But then farming itself held no certainties. The January, 1921, cover of *Agricultural Alberta* summed up the farmer's dilemma. "It's a great game but watch your move," warned the caption illustrating a farmer pondering a chess board. Each chess piece represented one of the many variables in agriculture. Help, a straw-hatted figure with a pitchfork, shared board space with Sunshine, Rain, Frost, Drought, Hail, and Price, all clearly as capricious as the farm worker.[48] Hired hands were unreliable, likely to quit their jobs just when they were most needed. And they were expensive. Farm workers demanded wages that made serious inroads into a farmer's often narrow margin of profit. Farmers felt themselves captive to a demanding and erratic labour force. But unlike weather and overseas markets, farm labour was one element over which farmers could exercise some control.

Broad attempts to solve the labour problems of unreliability and expense were only partly successful. Formal controls of the state and informal constraints of agrarian society worked to keep a lid on dissent, but they did little to solve the problem of an unreliable and expensive work force. Flooding the labour market was more effective in keeping wages down, but it did not solve the problem entirely, since the cheaper labour force did not have the necessary skills. Experienced men could still command wages farmers were unprepared to pay.

Perhaps the answer lay in technology. During the pioneering period of agricultural development, mechanization expanded production but did not reduce labour needs. But in the 1920s, farmers sought ways to reduce both their labour costs and their labour requirements, especially in light of the wartime experience of increased production despite labour shortages. During the First World War farmers had seen, on a limited scale, how mechanization and the reorganization of farm work could increase the productive capacity of the agricultural labour force. If technology could be used to a greater extent, the wage bill need no longer be seen as a necessary expense in order to improve production or to keep other costs down, but as a heavy cost itself that could be drastically reduced

Figure 3

Farm Operating Expenses, Prairies, 1927

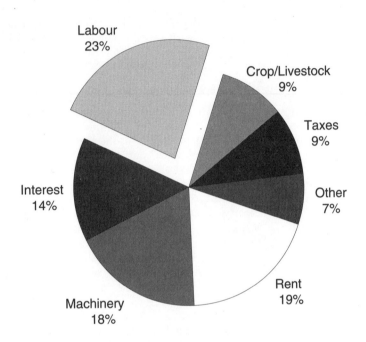

or even eliminated. Advertisements for farm machinery reminded farmers that "to make a profit, you must overcome not only handicaps of weather, weeds and pests, but also scarcity of help and high wages." Machinery such as the J.I. Case Threshing Machine could solve the problem. "The bright side of the picture is that your power and labor costs, which are under your control, can be greatly reduced, thereby giving you a profit year after year despite all handicaps."[49]

Changes in the techniques and organization of work have been subject to much recent debate on the nature of work.[50] Labour process theory has been revitalized in light of Harry Braverman's powerful thesis on relations between labour and capital with the maturation of monopoly capitalism. Marx's formulation of the relationship between the production process and social relations, especially the way that changes in the former resulted in comprehensive and extensive changes in the latter, provided the cornerstone for this work. Braverman found that management designed changes in the organization and technique of work in the twentieth century as a conscious strategy to remove control from the work force by fragmenting and deskilling its work. His thesis has been acclaimed as a welcome refocusing of attention to the workplace as the locus of labour-capital struggles. Recent and more detailed studies of changes in the labour process remain indebted to Braverman, but they have pointed out that the process and results of technological and organizational change were much more complex than he indicated. While it has been generally recognized that his crediting of a conscious management strategy and of complete labour subordination was greatly overstated, the complexities of changes in the work process are still being discovered.

In the case of prairie agriculture, the complexities are especially evident. In an industry that exercised little or no control over most aspects of its economics, the most effective way to increase profits was to increase production while reducing costs. It was a tall order. Agricultural output was subject to the ravages of pests and diseases and the uncertainty of weather. Costs were dictated by interest rates, tariffs, and transportation rates, none of which agricultural producers determined. Prices were beyond the farmer's ability to control, set by international market demand and local and world-wide production. One of the few variables that farmers might be able to regulate was the cost of labour.

In other industries, reducing labour costs can be achieved by manipulating the labour process. One avenue is to reduce the necessary skill level of individual workers by fragmenting and simplifying each stage of their work and thus being able to draw on a larger pool of lower-paid labour. Another is to intensify the productive capacity of each worker by speeding up every stage of the production process. Technology can play a significant part in achieving both these objectives. In the prairie West, however, farmers were warned that "factory methods of mass production cannot be applied to agriculture."[51] Although some highly skilled jobs were eliminated and others were speeded up, the main thrust of technology was to increase labour efficiency. This was not always to be achieved through deskilling and speeding up.

A significant difference between agriculture and other industries lies in the composition of its work force. In other industries, capital is counterposed to labour and the functions of each are clearly defined and separated. In agriculture, the two are juxtaposed: farmers and their hired hands work together in the production process. But because farmers can provide basic production themselves, their purpose in hiring additional labour is not to replace their own labour but to augment it. Since farmers themselves were providing the bulk of labour, technological innovation and mechanization were not designed to create a two-tier level of skill division. In other industries, labour's strength derived from its possession of special skills necessary for production, skills that management did not share, and skills that management had to appropriate in order to maintain control of its work force. But in agriculture, farmers sought hired hands with a skill level equal to their own. The effect on labour-capital relations was to establish the definition and possession of skill as bargaining tools.

The question of skill in agricultural labour had long been a point of contention between hired hands and their employers. Farmers in the Qu'Appelle district were asked in 1899 if they "sufficiently note the difference between the skilled and unskilled man all the way through?" Farmers had high expectations. "Our man should be able to handle a team carefully, be a good plowman, good on mower and binder and able to build a stack that will throw off rain," declared G.C.D. Edmunds in his address to the Farmers' Institute. "The good man is regular in his work, makes fewer breakages, manages his horses better and with fewer cuss words."[52] Twenty years later the question of skill was still not resolved. "While usually looked on as unskilled labour," reported the Alberta Employment Service in 1921, "the farm hand must have as much skill and experience in the handling of horses and machinery as the city tradesman or mechanic has in the use of the tools of his trade."[53] Although it was widely recognized that not all farm workers possessed the same degree of skill, there were no set criteria as to what constituted the necessary minimum. Charles More, in his study of *Skill and the English Working Class*, provides a definition of skill as "the alliance of manual skill with knowledge."[54] But in agriculture, the technical expertise that agricultural labourers possessed did not achieve a categorization according to industry-wide standards. Braverman points to the turn-of-the-century American census homogenization of all waged farm workers into one category of "farm laborers and foremen," arguing that "there is not even a hint . . . of an attempt to sort workers by skill."[55] He argues that this homogenization led to a disregard for the wide range of skills farm workers possessed.

Yet informal working knowledge was widely recognized, even if valued only in its absence, by prairie farmers and their hired hands. Ken Kusterer's application of the term *Know-How on the Job* provides a better understanding of the combination of proficiencies recognized as genuine skills with those that were essential but that passed unrecognized and unacclaimed.[56] This conceptualization explains the often contradictory regard and recognition among both hired hands and their employers for the degree of competence farm workers possessed. The widespread solution was not to label farm work as skilled, semi-skilled, and unskilled. Rather, the term most commonly used in agriculture to describe the most valuable men was "experienced."

The distinction was more than mere semantics. Although it included degrees of technical expertise, the term also encompassed judgement about farming conditions that could only be learned by dealing first-hand with the vagaries of nature. Prairie farmers and their hands recognized that experience meant more than simply the technical dexterity to perform farm tasks – knowledge of how to apply the manual skills required training. An experienced man was an all-round man, one who possessed a wide variety of skills that had been tested and refined through practical application. A good man should be able to harness a team of horses and operate the ploughs or harrows or seeders the team pulled, but this task was not simply a mechanical one. It required finesse in the difficult task of cajoling a recalcitrant team or manipulating tricky machinery over variable terrain. An experienced man had to be resourceful, to be able to make sound decisions about weather conditions or emergencies.

There was an additional qualification. An experienced prairie hired hand had to possess more than technical skill and working knowledge; he had to be willing not to be confined by a certain skill level. On a prairie farm, no task was too large or too small, too elevated or too mundane for the experienced worker. The "good all-round man" should be willing to spend a week picking rocks or repairing fences and just as readily turn his hand to slaughtering a hog or operating a tractor. An experienced hand was expected to know that a one-man operation as complex as a farm required the owner, and thus his hired hand, to be competent in every task.

With a designation so susceptible to differing expectations and interpretations, and so determining of wages, it is not surprising to find sharp differences between farmers and their hands over just what entitled a man to claim to be "experienced." Studies of the social construction of skill have emphasized that it is an area of continuing struggle between workers and employers.[57] When capital succeeds in providing the definition of different levels of proficiency, it is able to exert hierarchical control over its work force and to keep wage levels down. When labour is able to determine the definition, it is able to protect segments of its membership, to achieve some autonomy in the labour process, and to raise wages. Farm hands and their employers had strong opinions about what constituted experience and how it should be rewarded.

During the early settlement period, with a scarcity of labour and an abundance of land, farm workers held the upper hand in determining the criteria and rewards for experience. Near Wellington, Manitoba, in 1882, Arthur Sherwood was confident that even his limited skills could help him land a farm job: "I can milk which is a thing that talks out here."[58] Although farmers were constantly decrying the high cost and inefficiency of inexperienced workers, they often had little choice but to accept a prospective employee's own claims.

But by the 1920s, farmers could afford to be more stringent in their expectations. With the glut of unskilled workers on the labour market, farmers turned down men whom they even suspected of lacking appropriate experience. And even experienced men, unless they were well known in a district, were often forced to accept a lower wage for a month or two while they proved their abilities. Farm placement agents told of newcomers who "expected to get from $50.00 to $75.00 per month" and were "sorely disappointed" to receive only $25 because of their inexperience.[59] In the highly competitive agricultural world and the well-stocked labour market of the 1920s, farmers were unable and unwilling to allow their hired hands time to learn on the job. Farmers rationalized by concluding that it "was not fair to the man" to expect him to learn on the farmer's time.[60] But costs were the obvious explanation. A newcomer was "not worth over $25.00 per month to the Canadian farmer for the first few months," reported the Olds Colonization Board agent. "The farmer who gets him has to oversee and teach him and stand his breakages, etc., and he is really dear at $25.00 per month."[61] For short-term or seasonal jobs, the wage shortfall made a substantial difference in the final wage packet. Some men simply refused to accept these terms. When John Thomas and Gomer Jones, who had been expecting $60 per month, were offered $25 per month to begin and "a higher wage later in accordance with their ability, . . . they immediately left Olds [Alberta] and returned to Winnipeg."[62]

The value placed on farming skills was respected and undervalued at the same time. From their own experience, farmers knew exactly what was entailed in every aspect of farm work. In placing high value on their own skills, they were giving implicit recognition to the value of their hired hands' expertise. This could lead to extremely stringent, even unattainable, criteria by those farmers who had a very high regard for their own abilities.

In principle, too, it could lead to a very high wage. This was seldom the case. Farm hands were rewarded for their experience and could sometimes double the wage of inexperienced men. But the overall low wages in agriculture undercut this advantage. Compared to the rate of difference between much less-skilled harvesters and regular hired hands, it was clear that scarcity of labour was much more powerful in determining wage rates than was skill or experience. In fact, unless a regular experienced hand was given a bonus or put onto harvest rates, his wage during harvest was likely to be lower than that of an inexperienced stooker.

Traditionally, farm workers had accepted the relatively low monetary reward for their expertise because they would not simply be attempting to sell it in an open marketplace but would use it eventually in their own enterprises. Their payoff was not in the wages they received but in the experience they gained, which would enable them to succeed on their own farms. Thus the opportunity to acquire farming skills or experience was likewise a point of contention between farmers and their hands and an arena of continuing struggle. The apprenticeship aspects of farm labour were a powerful draw to men who had no agricultural background and who were eager to learn farming skills. Noel Copping explained the complexity of the learning process:

> [I] am gradually becoming more proficient at the art of driving a sulky plough – and it is an art: when you set in to start a furrow there is a lever to press down with the foot and which is locked by a hand lever. Then there are other levers – two, one on the left and the other on the right, that have to be manipulated, and there are also the reins to hold, and the whip. It can therefore be seen that at least half a dozen hands are necessary. I have only two! Then too, one has to keep one's eyes on the four horses, on the furrow wheel, on the sods behind in case they fall back and on everything else.[63]

A particular problem for inexperienced men was that most of the skills required in farming could be picked up informally. Men from a rural background had a definite edge. Percy Maxwell was dismayed to discover the disadvantage at which his urban background placed him:

> It is also a bit sickening, when you think you are going to do great things such as ploughing, etc. and be a real farmer, to find that all the kids in the country have been doing the same things almost since they could walk. A kid of 14 here does a man's work and is expected to earn his own living.[64]

Even the simplest of farm jobs required the uninitiated to spend some time in learning, although they appeared as simple common tasks to those with a rural background. As Kusterer has pointed out, "the use of the 'unskilled' label has led to a gross under-estimation of the amount of working knowledge actually necessary in these jobs. There is no such thing as unskilled work."[65]

As a management strategy for reducing labour costs, fragmentation and deskilling had little place on the small farms that provided employment for the majority of hired hands. Farmers expected their hands to possess at least rudimentary familiarity with agriculture, and they much preferred to hire experienced men. The dispute between farmers and their hands narrowed down to one of the length of time required for learning. The promotion to a higher wage in recognition of expertise and experience was less easily achieved when farmers could draw from a larger pool of labour. During the early period, farmers were forced to accept men who were "willing" for men who were "experienced," but since experienced men could command a higher wage, newcomers were anxious to have their developing skills recognized. Still, both farmers and their hands agreed that the most valuable hired hand was one who possessed more, rather than less, skill.

Since reduction of labour costs could not be easily achieved through simplification of the labour process, the only alternative was to reduce labour requirements. The major agricultural developments of the decade – technological innovations and the slow and intermittent increase in mechanization, the growing specialization in wheat, and the consolidation of farm lands – combined to secure this result. Although not all of these developments were planned as labour-reducing economies, they either directly or indirectly achieved that effect.

Mechanization is a case in point. Farmers had always complained about the scarcity of farm help. But the "labour shortage" of the 1920s was not absolute, as it had been during much of the pre-war period. Rather, it was a relative shortage resulting from the refusal of men to work for low wages. The thrust of technological improvements and mechanization in the 1920s reflected this situation. Although mechanization in prairie agriculture was still in its infancy, important patterns in its purpose were emerging. During the pre-war period most technological innovations had been designed primarily to increase production, even if this resulted in greater labour requirements. Crews of eighteen or twenty men were needed in steam threshing operations, but they were essential to the expansion of wheat production after the turn of the century. During the 1920s, innovations were still intended to expand production, but the overall purpose was to make the individual family farm self-sufficient with regard to labour. When the giant wheat farm operated by the Scottish Co-operative Wholesale Society at Hughton, Saskatchewan, "passed out of existence" in 1926, the *Grain Growers' Guide* editorialized that "industrial methods cannot be applied to agriculture." Farmers were encouraged to adopt business-like methods of farming, but cautioned against travelling too far down that path. "The future of this country," declared the *Guide*, "depends largely on maintaining a class of free and independent farmers, owning their own farms and running their own business."[66] In the context of world markets and prairie farm debt, this meant a continuing increase in production combined with a relative reduction in labour requirements.[67] The ultimate imperative was to reduce labour costs by reducing or even eliminating the need for waged labour.

The trend toward mechanization during the 1920s was thus a response of the agricultural industry to a labour force that refused to work for low wages. "Labor conditions make it necessary and profitable to invest in mechanical contrivances that take the place of manpower," declared the *Nor'-West Farmer* in 1920, when post-war labour shortages were still acute and wages were still high.[68] It was a circular process. By demanding decent wages, farm workers forced the industry to find alternate methods to increase production. The search for technological improvements and the resulting mechanization led ultimately to a shrinking of labour needs. But it was a lengthy process and, as the major component in the drive toward rationalization in the industry, it achieved uneven success. For the most part, mechanization did not bring about an immediate drop in labour requirements. The labour saved both by minor technical improvements to agricultural implements and by major technological advances in machinery and equipment resulted not in a reduction of labour but in greater labour efficiency.

Technological changes in production methods increased agricultural output. At the beginning of the decade, steam power and mechanization were already used to speed up the most time-pressured farm work. Ploughing, harvesting, and threshing could all be done with steam power and machinery. Throughout the 1920s, continued improvements were made to farm implements to increase their efficiency, taking advantage of the "progress in the industrial arts" that was shaping other industries.[69] But the most significant changes during the decade were a shift in the source of power and the growing use of more complex machinery.[70] When the *Canadian Thresherman and Farmer* changed its name to *The Canadian Power Farmer*, it declared that "the individual farmer is an engineer and uses mechanical power on his farm to an extent that no-one probably dreamt of in 1902."[71] During the 1920s the agricultural industry began to adopt three pieces of equipment that had already seen widespread use in the grain fields of the western United States: the gasoline tractor, the small portable thresher, and the combination harvester-thresher.

Although gasoline tractors had been available since the 1890s, they did not gain widespread acceptance on prairie farms until the 1920s, and even then their rate of adoption was slow and sporadic. The early gasoline tractors had little to recommend them – they were unreliable in the field, expensive to buy and maintain, complex to operate and repair, and too powerful for farm implements designed for the power capabilities of horses. The introduction of the gasoline tractor was not uniform throughout the prairies, nor did it represent a complete switch from other sources of power. Steam power continued to be used in threshing operations, but it was gradually phased out as the superior efficiency of the gasoline engine was demonstrated.

Horse power was much more slowly given up. Even farmers who did buy tractors seldom gave up all their horses at once. The debate on the relative costs and merits of horses against gasoline tractors was

Farm workers' skills evolved with changing technology. (Provincial Archives of Manitoba, Agriculture – Machinery 24, c. 1910.)

carried on throughout the decade. Expert opinion held that tractors were cost-efficient only on large operations. On small farms, tractors were considered a worthwhile investment only when the prices of feed and labour were high, but it was never easy to compute the actual relative costs.[72] Tractors represented a high fixed cost, both in purchase price and in running and maintenance, and the work they could perform was restricted by topography and weather. Horses were much more flexible in all regards. They could reproduce themselves, they could subsist on the by-products of the farm, and their labour was adaptable to all farming terrain and conditions. A study by Andrew Stewart at the Manitoba Agricultural College in the early 1930s found it was necessary to operate a tractor for 600 hours per year to make it pay. The hours of work required for any particular task varied greatly across the prairies and even from farm to farm, depending on soil conditions and amount of bush, and were restricted by the Plains topography. When translated into the size of cropland necessary to give a tractor the requisite 600 hours of work per year, Stewart's figures ranged from 282 acres to 857 acres.[73] Since improved acreage on prairie farms during the 1920s averaged about 50 per cent of the total area occupied, it was clearly necessary to consider purchase of a tractor in conjunction with the possibility of increasing farm size and increasing specialization in field crops.

The tractor was intended to replace the power of horses rather than of men and in doing so not to eliminate labour but to make more efficient use of it. Although horses could perform tasks that machines could not, machines and the men who operated them could be worked for much longer periods. The tractor reduced a farmer's manpower needs as well, by freeing him from the restrictions necessarily imposed by animal power. In part the shift resulted from a freeing of land and labour from the feeding and care of horses, but in part it was due to the greater labour efficiency of a man operating a tractor instead of handling a team of horses. The overall result was an increase in productivity. Stewart's study concluded that even in the brief period of their use, "tractors adopted on the farms of the Prairie Provinces seem to have added to the acreage of field crops at the rate of about 1,000,000 acres per year."[74] Most of the change came about in the second half of the decade. After 1926, the volume of tractor sales and the acres of field crops both rose, while the numbers of horses began to decline.[75]

Tractors were used mainly in the spring and fall, the peak times of labour need. In the spring they pulled ploughs and harrows to prepare fields for planting, and in the fall they pulled power binders or combines and provided the belt power for threshing. The great advantage of tractors was in the speed with which these operations could be completed and in the flexibility of labour needs. Both had the indirect result of reducing labour needs by allowing greater production in a shorter time. Reducing the man-to-land ratio resulted in greater profits. Explaining the growth of farm size as the impetus behind the necessity for speeding up farm operations, J.G. Taggart, superintendent of the Dominion Experimental Farm at Swift Current, pointed out that the "increased acreage is being handled without anything like a corresponding increase in personnel [which] makes speed in the performance of both seeding and harvesting operations most essential." Taggart explained the indirect savings in labour costs:

> A study of the relation of the tractor to the evolution of farm practice clearly indicates that the reason for the fairly extensive use of a tractor is that each man can handle a larger acreage, rather than effecting any great saving in the cost per acre of specific operations. It is obvious that if a man can seed and harvest four hundred acres instead of three hundred in the available season for such work, then he will gain by the amount of the net return from the additional one hundred acres.[76]

In 1928, prairie farmers bought more than 6,000 small threshing machines to be run by the belt power of gasoline tractors,[77] thus freeing themselves from the expense and time constraints of the large custom steam threshers.

Although the cost of labour was by no means the only consideration in a farmer's decision to buy a tractor, it was an important one. The surge in tractor sales at the end of World War One was fuelled in part by escalating farm wages, which peaked in 1920 at more than three-and-a-half times the pre-war figure.[78] A drop in wages the next year was at least partly responsible for the subsequent decline in sales. By the end of the decade, though, the technology had improved and purchase costs had fallen enough to make the gasoline tractor seem a sound investment.

Farmers throughout the 1920s were faced with a barrage of advice to improve their farming efficiency. A decade of advertisements extolling the virtues of machines over hired hands may well have induced farmers to take the plunge. "When money is paid to hired help it is gone," declared the *Nor'-West Farmer*: "it is true you have had the help – but nothing else. When money is put into labor-saving machinery, the machinery remains, to give help and save time and labor for the rest of your life. The interest on the investment is repaid over and over again."[79] Farmers were ready, once they brought in a good crop in 1927 or 1928, to buy a tractor to reduce their dependency on waged labour. After "Fifteen Years of Tractor Progress," the International Harvester Company declared its Titan 10-20 Kerosene Tractor so simple to operate that "a fourteen-year-old boy can handle the Titan and do a man's work."[80]

The difficulty of keeping reliable farm workers was instrumental in increasing tractor sales at the end of the decade. Farmers were urged to solve their labour needs with machinery and technology. "Relieve Shortage of Farm Help" exclaimed the advertisement for Case Power Farming Machinery.[81] The Litscher Lite Plant was "A Hired Man You Can Keep."[82] Farmers looking to solve their problems of labour shortage while increasing their production were told that "the tractor is the only solution for a man who wants to farm a fair sized or a large farm without outside help."[83] The availability of grain separators with their concomitant reduction of labour provided even more incentive to purchase tractors. Sales mushroomed in 1927 and remained high until the collapse of grain prices in 1929.[84]

The savings in labour costs and convenience with the use of tractor power for threshing made the introduction of the other major step toward mechanization, the combination harvester-thresher, even more welcome. Here the results on labour needs were immediate and direct. The combine, as it was commonly known, was introduced in the American West in the 1880s, but it did not gain acceptance on the Canadian prairies until the late 1920s, when its use could be adapted to the shorter prairie growing season. Once swathing was developed, the combine was adopted. Swathing allowed the grain to be cut while it was still green, then laid on the stubble to dry and ripen until it could be collected for threshing. This reduced the dangers to a standing crop of pests or hail or an early frost.[85] The tractor was necessary to provide the power, and sales of combines reflected the spurt in tractor sales in 1928 and 1929.[86]

As the technique was adopted, the demand for harvest labour was dramatically reduced. One study of combine use showed a reduction in man-labour hours per acre "from 4.6 hours for binding, [stooking] and threshing with a stationary thresher, to 0.75 hours for work with a combine."[87] Combines did not come into widespread use on the prairies until after the Second World War, but their appearance in the late 1920s was enough to cause satisfaction among farmers, for whom "the elimination of extra help in harvest at high wages has been long looked for."[88] Farmers calculated the costs and found the combine a worthwhile investment. Farmer Anthony Tyson kept accounts of his harvesting expenses for the three years between 1927 and 1929, when wages were relatively steady. He spent $2,897.62 and compared this to the $1,549.57 he would have spent if he had bought a combine and done the harvesting himself. His costs for labour were $1,369, which "would have been halved" if he had purchased a combine.[89] But the advent of combines caused some alarm in the Department of Labour, which calculated that "every combine deprives at least 5 men of a harvest job."[90] For hired hands, the change was unsettling. When Ebe

Koeppen went to the area around Calgary to "make some money on the harvest," he found that "combines were already coming in, so there was very little work around."[91]

Mechanization proceeded slowly on the prairies, and in the post-war decade it was still in its infancy. Nonetheless, it did have a significant effect on labour requirements, particularly since the farmers with the greatest labour needs were the ones who bought combines and other labour-reducing technologies. When farmers did move toward mechanization, it was more often a gradual process than an abrupt switch.

For hired hands, too, the change came subtly. The seasonal pattern of agricultural employment meant that farm workers seldom faced outright displacement by machinery. Rather, men who had found work for a specific task in one year might find that by the following year their employer had mechanized his operations and no longer needed their labour. A farmer who bought a combine, for example, needed only two or three men for his next harvest season instead of eighteen or twenty. A farmer who bought a tractor might still hire a man to operate it but would need him for only a week instead of two.

Mechanization did reduce overall labour requirements, but at the same time it changed the type of labour needs, providing greater opportunity for steady work, as a 1930 Alberta Department of Labour Report details:

> In the older settled wheat growing sections of the Province, tractors and other machinery have been introduced to such an extent that a complete change in the system has taken place which has brought about a material reduction in the number of men required to handle spring and summer work, and along with this has come more stable employment for men on the farms. Quite a number of farmers, who in past years employed a volume of help in the spring, releasing the men following the spring operations and then hired an additional volume for harvest work, now employ a small number of men steady the year round on their farms, and as a result of mechanizing do not require any additional men during any period of the year.[92]

There is no doubt that mechanization resulted in increased agricultural productivity, but it did so at the cost of placing the machinery, rather than the farmer and his men and animals, at the centre of the production process. In the early years of the century, harvesting the prairie crops could take several weeks, and threshing operations could be spread over a period of five or six months. Where combines were introduced in the late 1920s, the time was considerably reduced. Even the bumper crop of 1928 was harvested and threshed in only three months. Moreover, the combine "altered the farmer's work year drastically," by "increasing the need to plant early so as to insure a ripe crop in time for combining."[93]

Mechanization was also closely bound up with two other trends of the decade: growing farm size and increased agricultural specialization. It is difficult to distinguish cause and effect. The increasing mechanization and the development of new technologies allowed more and more acreage to be brought under cultivation, and the consolidation of farms into ever larger units enabled the most efficient use of the new machinery. But machines were expensive, and since the high capital investment demanded their most efficient use, farmers had to increase the size of their holdings to justify the expense. The same interrelationship exists with regard to specialization. Mechanization allowed economies of scale that were most readily realized in crop specialization, and once the investments were made it became more difficult to diversify, something that prairie farmers were reluctant to do in any case. At the same time, the reduction in the use of horses, both on the farms and in the cities, freed large acreages for wheat production that had previously been used for grazing or raising feed.

For the new technology to pay for itself, farming operations had to be reorganized and geared to the machine. This development dovetailed neatly with the increasing emphasis on efficiency. Mechanization facilitated the trend to more business-like operations that characterized agriculture during the 1920s.

For farm workers, the changes resulting from mechanization were complex, and the immediate benefits often outweighed the disadvantages. Where the capacity of a machine dictated a worker's performance, the change was often a trade-off of one task for another. A tractor might tie a man to the fields for a twelve- or fifteen-hour day instead of the ten hours that could be expected of a team of horses, but the work of harnessing and unharnessing the horses, and feeding and grooming them after the working day, ate up the extra time.

Machines also eliminated some of the most back-breaking labour and drudgery from farm tasks. The romantic picture of a farm hand following a horse-drawn implement leisurely around a field is belied by the description left by Gaston Giscard: "the harrow hasn't got a seat on it, which makes it a tiring job. It wears you out because you're walking all day, swallowing the dust thrown up by the horses."[94] The same job on a tractor might be just as monotonous, but it was less physically demanding and more quickly finished.

For farm workers who preferred a slower pace, there was still plenty of opportunity to practice traditional agricultural methods. Technological advances and mechanization did not occur at a uniform rate throughout the prairies. As in most other industries, the rate of change was uneven and irregular, with the most advanced machinery being operated beside equipment whose design had not changed in decades. The differences were even more striking in agriculture. With thousands of individual farm enterprises, each instituting change at its own pace, a hired hand could find himself working under the most advanced or the most primitive conditions. When W.N. Rolfe moved from the farm of Reg Matthews to that of Percy Scharf, he found "a complete change in every way. Here was pioneer western farming of the first order, rough and heavy and varied as it is possible to find it."[95]

It is not surprising that farm workers greeted the new technology with equanimity. The thrust of mechanization was an attempt to increase the productive capacity of workers but not through pace-setting or skill-reducing technology. Labour in prairie agriculture was thus not faced with the problem of skill dilution. In some cases, entirely new and higher levels of skill were called for. At a time when the College of Agriculture at Saskatoon was offering three-week courses on learning to run a tractor,[96] the man who could operate one was highly prized.

Overall, the new technologies resulted in a complex process of reskilling. Every year brought innovations and improvements to agricultural technology, yet because the changes were irregular and incomplete, farm workers became adept at maintaining old skills while learning new ones. Hired hands who had wrestled with the intricacies of the harness were now learning how to diagnose and repair a faulty carburetor. When farmer A. Schoonover was looking for a farm hand, he specified that he wanted "a man who can handle horses" and who "must understand farm machinery."[97] By the 1920s a "good all-round hand" was one who could master the complexity of the internal combustion engine and still know how to treat the internal apparatus of a colicky horse.

Machinery did not remove autonomy. Even while making use of the new agricultural equipment, farm workers maintained considerable control over the direction and pace of their work. Mechanization in agriculture involved much more than simple machine-tending. Mechanization was giving rise to "new standards of labor," according to one rural sociologist. "Hence the human ox is a fallen idol, and the man of wits is being exalted."[98] Hired hands' mechanical abilities had to include dealing with the idiosyncrasies of the new machinery, which might break down in the field, had to be constantly adjusted to the changing physical conditions, and required skill and dexterity to make it produce to its full capacity.

Machines thus presented new challenges, and rather than being regarded as harbingers of future unemployment or dictatorial automatons, they were seen as technology to be mastered. Unlike farm workers on the Argentine pampas or in Ontario or the American Midwest in an earlier period,[99] prairie farm workers did not fight to prevent the arrival of the new technology. Instead, they relied on their ingenuity and resourcefulness to harness it. On a personal level, a hired man could achieve as much satisfaction from learning how to coax new life from an ailing engine as he had from learning how to trick a balky pair of oxen into pulling a load through a muddy creek. In his relationship with his employer, he could parlay his newly acquired skills into higher wages. But in the larger sphere of labour-capital relations in the agricultural industry, machinery was only one weapon that both labour and capital could use in the ongoing struggle.

Farm workers in the 1920s were caught up in the advancement of capitalist agriculture. Their proletarianization was defined by their increasing difficulty in escaping from wage labour to farm ownership. In their attempts to maintain the social and economic advantages they had previously enjoyed, they clung to the independence that had served them well in the earlier period. Individualism continued to be both a structural and cultural condition of their experience, and hired hands appropriated and reshaped parts of the dominant ideology to suit their needs and to mitigate the increasing tensions in the social sphere.

But in the workplace they fought back. Here, too, independence structurally and culturally conditioned farm workers' experiences and their responses. They resisted the increasing advantage of capital in a daily and individualistic struggle. In the aggregate, their actions added up to a powerful obstruction for the agricultural industry. Capital responded by using mechanization and technology to overcome an obdurate work force. Labour responded by using newly acquired skills as a bargaining tool. The workplace continued to be "contested terrain," where labour and capital worked out an uneasy relationship in which neither side was the clear winner.

9

Conclusion

"I think Frank was glad of a permanent home," declared a farmer near Three Hills, Alberta, "and being practically his own boss as far as the work was concerned on the farm."[1] Frank Ready was a "professional hired man," according to his neighbours,[2] and his apparent satisfaction derived both from his inclusion in the farming community and from the independence possible in his labour. This has made his role problematic for labour historians and helps to explain why hired hands have been excluded from accounts examining the growing class consciousness of Canadian workers. It also helps to explain why farm workers have been submerged in the more heroic accounts of pioneering and the development of the agricultural industry. Despite their numbers and their important role in the prairie West, hired hands have remained shadowy figures, and their story has more often been told by others than by themselves.

Hired hands worked and lived in ways that showed little surface change during the period when the agricultural economy of western Canada was developing and expanding. Throughout this half-century, they received meagre returns for work that was onerous and dangerous. Employment was irregular, subject to seasonal and economic cycles. Wages were low and difficult to collect. Hours were longer than in other industries, and tasks were often tedious and repetitive. Ameliorative labour legislation specifically excluded farm workers, who received no legal protection beyond the Masters and Servants Acts in each province. Living conditions were likewise very poor, or at best adequate. Although many farmers boasted that they treated hired hands like members of the family, many farm workers complained of filthy conditions and food not fit for animals. Always, the living conditions of hired hands depended on the wealth and generosity of their employers.

However, there were discontinuities as well, as the underlying conditions of farm workers' productive and social relations were transformed. During the settlement period farm workers were part of the pioneering process. They entered the agricultural community as anticipated peers and were frequently accorded the same social status as men who owned their own farms. Indeed, many farm hands were at the same time farmers themselves, homesteaders who worked on another farm for six months a year to gain a stake to last through the winter and buy seed and equipment in the spring. Their social status was high and assured, and their living standards frequently no worse than those of the pioneer farmers for whom they worked. When farmers were struggling to create farms with very little capital, there were few distinctions between employer and employee in living or working conditions.

But as farming communities passed from the pioneering stages into the more prosperous, businesslike atmosphere of the First World War and the 1920s, the lives of farm workers remained unchanged. In this respect, then, they suffered a relative deterioration of their living and working conditions. As farm homes became cleaner, more genteel places for farm families, the hired hand was less and less welcomed into the household. Although the living standards of farmers improved, those of farm workers frequently remained in the same crude mould in which they had been cast during the pioneering period. Some hired hands did enjoy the improved living standard of their employers, but many others were still relegated to abandoned sod huts, chicken coops, or granaries.

Working standards remained low for many farm workers, too. Those who learned to operate and repair machinery were in high demand, but the growing numbers of unskilled workers, whose ranks were swollen as a result of immigration policies in the 1920s and increased mechanization, were faced with the same dangerous and onerous tasks that unskilled workers had always performed. There was little opportunity to cut the tedium by other more specialized farm tasks, and little chance to improve their skills because farmers no longer had the time, or needed to take it in a more abundant labour market, to teach farming techniques to newcomers.

The most significant change in the lives of hired hands in the 1920s, though few of them realized it, was the breakdown of the agricultural ladder. The hopes for future independence that had motivated agricultural labourers in the period before the First World War continued to check worker discontent, but these hopes resulted in subtle differences in the social position of farm workers during the twenties. Since they had little real chance of becoming farm owners themselves, their social position deteriorated. Although still welcomed at social functions such as dances and picnics, they could seldom any longer aspire to real entry into a farming community that was becoming more business-oriented, with an increasing drive for rationalization of the industry, increased political and economic power, and a distancing between employers and employees.

Yet farm hands frequently accepted these conditions with very little complaint. They were individualists who relied on their own resources. Hardship was part of the bargain of farming, and adversities could best be overcome by exercising individual initiative. Many hired hands continued to hope that they would soon leave the ranks of wage labour and become farmers themselves. And there were external constraints as well. Severe government repression of radical labour organizations such as the Industrial Workers of the World, which attempted to organize farm workers, kept discontent silent. Close and continuous contact with their employers encouraged hired hands to regard their grievances as personal ones between their employers and themselves rather than as conditions within the industry or the capitalist system. Their isolation on widely scattered farms encouraged self-reliance and fostered the spirit of individualism that drew farm hands to the work in the first place. During most of this period, farm workers could take advantage of a tactic that gave them independence in their working lives and a degree of control in their relationship with capital. If they were unhappy, they could vote with their feet. Farm workers chose this route in the early pioneering days and were still following it fifty years later.

Men undertook agricultural labour for a variety of reasons, but the common thread running through their choices was independence. Whatever penchant hired hands may have had for an individualistic outlook was reinforced when they were confronted with the realities of their economic position. Individualism gave them their greatest weapon and provided them the greatest cushion against the conditions of their proletarianization. In turn, they embraced individualism to an even greater extent, incorporating it into their culture of survival and resistance, both in the social setting and in the workplace.

Yet it would place too great a burden on the evidence to claim that hired hands developed an independent culture of their own. Through their physical and social isolation from other farm workers, through their acceptance into the dominant agrarian culture, and through their hopes of entering the ranks of farm owners, hired hands were largely absorbed into the cultural world of their employers. However, in their common responses to the conditions of their lives and labour, they demonstrated a kinship with other hired hands and with the larger cultural world of workers in other industries. Farm workers occupied an important position in the spectrum of the Canadian labour experience.

This study of labour-capital relations in prairie agriculture thus has broader applicability, for the experiences of hired hands on isolated prairie farms tell us much about how individual workers have dealt with the realities of their proletarianization. Their first response was to operate within the relationship as defined by capital. They used the informal apprenticeship system to develop farming expertise, to raise capital, and to make their way into the agrarian community. Yet they also took advantage of the availability of land and the shortage of labour to strike bargains over wages and working conditions. And they developed skills to reap advantages from the close working and living relationships with their employers. Hired hands had necessarily to extend their theory and practice of labour-capital relations well beyond the simple antagonisms of class. Much more than a simple acquiescence, a nuanced understanding of personal class relations enabled farm hands to operate effectively even without recourse to the usual institutions of working-class support such as unions or community.

Another response was to subvert underlying features of the relationship. Their own interpretations of cultural constructs such as bachelorhood and masculinity gave hired hands a sense of their own identity. This provided a support against the increasing divergence of interests between labour and capital, and against the more conflictual nature of the relationship. Their role in transformations of the labour process gave them a degree of control over important aspects of their working lives. They embraced those techniques and devices that provided relief from back-breaking labour, yet fought to retain those aspects of the work they found rewarding. They developed skills and technical expertise to supplement the informal "know-how" of the job in order to maintain the control that figured so largely in reducing much of the alienation inherent in waged labour.

Ultimately, the response of farm workers was to resist the conditions of their proletarianization with the most potent weapon they had – withholding labour. They used this tactic during periods of labour shortages to drive up wages. The more highly skilled hands used it even during periods of relative labour surplus to regulate the conditions of their work. And all farm workers used it as the ultimate protest against working or living conditions. On an individual level, this tactic caused grief for their employers and a measure of control for themselves. In the aggregate it forced capital to seek new ways to deal with the persistent problem of labour.

But job-jumping among hired hands, like other forms of their resistance, was not a consciously collective strategy. As a result, farm workers have been excluded from studies of working-class formation and struggle. Yet despite the particularities of early prairie agriculture that shaped labour-capital relations, the actions of farm workers had parallels in other industries. Although they remained outside the mainstream of labour organizing, hired hands exercised tactics that added up to a pragmatic response to their specific conditions yet were common to workers everywhere.

From the early pioneer period until the onset of the Great Depression, hired hands in the prairie West worked out their relations with capital in the context of choice. Farm workers found their courses of action increasingly restricted as the industry moved toward capitalist agriculture, but they nonetheless continued to operate on their own agenda. Far from colluding in their own oppression, prairie farm workers displayed agency. In their lives and in their labour, they demonstrated an ingenuity and a resilience that gave them a unique perspective on relations between labour and capital, yet assured them a firm position within the Canadian working class.

Appendix

Table 1

Agricultural Work Force

	Paid Workers	Farmers' Sons	Farmers	Total Agricultural Work Force
A. Manitoba				
1881	—	2,608	10,818	13,565
1891	4,981	5,353	20,322	30,988
1911	16,815	6,394	44,139	68,654
1921	16,593	14,875	50,845	83,229
1931	19,624	22,750	47,371	90,761
B. Saskatchewan (includes NWT before 1905)				
1881	—	180	822	1,051
1891	1,337	1,870	7,691	11,861
1911	20,097	10,090	99,510	130,983
1921	30,314	22,662	113,424	167,294
1931	37,221	47,488	119,799	200,236
C. Alberta (included in Saskatchewan before 1905)				
1911	7,990	5,582	61,599	79,067
1921	16,351	13,619	79,004	110,911
1931	25,616	26,194	89,686	142,328
D. Prairies				
1881	—	2,788	11,640	14,616
1891	6,318	7,223	28,013	42,849
1911	44,902	22,066	205,248	278,704
1921	63,258	51,156	243,273	361,434
1931	82,461	96,432	256,856	433,325

NOTE: Data are for males, 15 years and over, except for 1921, which is for males, 16 years and over. The census of 1901 did not include statistics on the agricultural work force.

SOURCES: *Census of Canada*, 1931, Vol. VII, Table 40, pp. 134-35, 146-47, 167-77; *Census of Canada*, 1921, Vol. IV, Table 4, pp. 242-45, 270-71, 292-93; *Census of Canada*, 1911, Vol. V, Table V, pp. 52-53, 96-97, 228-29; *Census of Canada*, 1891, Vol. II, Table XII, pp. 145-46, 181; *Census of Canada*, 1881, Vol. II, Table XIV, pp. 319-20, 327.

Table 2

Farm Workers' Wages

A. Annual Farm Workers' Wages
(not including board)

	Can.	Man.	Sask.	Alta.	Prairies
1901	$208	$155-$232	$150-$264		
1909	216	234	198	242	$224
1910	198	224	235	216	225
1914	152	179	168	168	172
1915	166	208	185	201	198
1916	194	237	215	267	240
1917	383	437	458	508	468
1918	429	515	549	551	538
1919	476	577	576	640	598
1920	543	650	667	697	671
1921	421	503	498	463	488
1922	359	391	413	367	390
1923	372	400	400	432	411
1924	380	365	407	389	387
1925	383	357	396	421	391
1926	384	352	404	422	393
1927	396	358	415	446	406
1928	382	353	411	450	405
1929	373	352	398	404	385
1930	326	298	340	342	327

SOURCES: *Canada Yearbook* (*CYB*), 1906, p. 159; *CYB*, 1914, p. 203; *CYB*, 1915, p. 176; *CYB*, 1916-17, p. 202; *CYB*, 1921, pp. 237-38; *CYB*, 1930, pp. 240-41; *CYB*, 1931, pp. 250-51. Provincial figures for 1901 are averages, calculated from *Labour Gazette*, vol. I, pp. 564-65.

Table 2 – cont.

B. Summer Farm Workers' Wages
(monthly, not including board)

	Can.	Man.	Sask.	Alta.	Prairies
1901	—	$18-24		$17-27	
1909	$24	25	22	25	$24
1910	23	25	26	23	25
1914	21	24	24	24	24
1915	23	30	25	27	27
1916	26	30	30	33	31
1917	45	47	50	53	50
1918	49	55	61	60	59
1919	54	63	66	67	65
1920	60	70	72	76	73
1921	45	53	54	52	53
1922	38	40	40	41	40
1923	40	40	42	46	43
1924	40	37	43	42	41
1925	40	38	42	44	41
1926	41	38	43	45	42
1927	40	38	43	45	42
1928	40	38	44	46	43
1929	40	38	44	43	42
1930	34	32	37	37	35

SOURCES: 1901, averages, calculated from *Labour Gazette*, Vol. I, pp. 564-65; *CYB*, 1915, p. 176; *CYB*, 1916-17, p. 202; *CYB*, 1921, pp. 237-38; *Labour Gazette*, 1930, Supplement to Jan., 1930, p. 100; *CYB*, 1931, pp. 250-51.

Table 2 – cont.

C. Annual Value of Farm Workers' Board

	Can.	Man.	Sask.	Alta.
1909	$120	$132	$192	$180
1910	150	176	168	200
1914	171	186	198	196
1915	175	183	201	203
1916	203	218	218	234
1917	228	252	276	276
1918	252	276	300	312
1919	288	312	336	336
1920	278	325	336	341
1921	248	295	297	283
1922	235	—	—	—
1923	239	—	—	—
1924	256	—	—	—
1925	258	260	268	280
1926	255	—	—	—
1927	262	254	277	290
1928	252	258	284	295
1929	254	256	287	274
1930	233	238	253	256

SOURCES: *CYB*, 1914, p. 203; *CYB*, 1915, p. 176; CYB, 1916-17, p. 202; *CYB*, 1921, p. 238 (the resulting figures for 1917-19 appear to be slightly high); *CYB*, 1930, pp. 240-41; *CYB*, 1931, pp. 250-51.

Table 2 – cont.

D. Saskatchewan Farm Workers' Wages
(monthly, not including board)

	Summer	*Yearly*
1907	—	$23
1908	$29	22
1912	31-40	23-29
1913	32-41	22.50-28.50
1914	25-35	20-25.50
1915	40-45	25-30
1916	50-55	37.50-40
1917	55-65	40-46
1918	55-65	50-54
1919		

	Winter	*Spring*	*Season*
1920		$60	
1921			
1922			
1923		50-60	$35-$45
1924	$10-$25	35-55	
1925	10-25	35-55	
1926	10-25	35-60	40-50
1927	10-25	35-50	50
1928	10-25	35-50	50
1929	10-25	35-50	40
1930	10-15	25-40	15-20
1931	board-10	5-25	board-10

SOURCES: Sask. Agric. report, 1907, p. 108; *ibid.*, 1909, p. 83; *ibid.*, 1919, p. 30; Sask. Labour report, 1920-21, p. 35; *ibid.*, 1924, p. 56; *ibid.*, 1925, p. 36; *ibid.*, 1926, p. 30; *ibid.*, 1927, p. 37; *ibid.*, 1928, p. 34; *ibid.*, 1929, p. 74; *ibid.*, 1930, p. 42; *ibid.*, 1931, p. 59; *ibid.*, 1932, p. 36.

Table 2 – cont.

E. Harvest wages
($ per day)

	Sask. figures	Prairie estimates
1901		1.88
1902		2.75
1903		2.00
1904		2.00
1905		2.25
1906		2.57
1907	1.50-2.50	2.00
1908	1.75-3.25	2.05
1909	2.50	2.00
1910		3.13
1911	2.50	2.88
1912	2.75-4.00	3.13
1913	2.75-3.75	3.13
1914	2.50	2.55
1915		2.60
1916	3.00	2.75
1917	3.50-4.50	4.00
1918	4.50	4.55
1919		4.69
1920		5.73
1921		3.88
1922		3.55
1923	3.50-6.00	3.75
1924	3.00-6.00	3.38
1925	3.50-5.00	4.10
1926	4.00-6.00	3.40
1927	4.00-6.00	4.50
1928	4.00-6.00	3.90
1929	2.50-4.00	3.58
1930	1.50-2.50	
1931	1.00-2.50	

SOURCES: Sask. figures: Saskatchewan Agriculture Report, 1908, p. 93; *ibid.*, 1909, p. 82; *ibid.*, 1910, p. 77; *ibid.*, 1912, p. 137; *ibid.*, 1913, p. 196; *ibid.*, 1918, p. 153; *ibid.*, 1919, p. 30. Prairie estimates: John H. Thompson, "Bringing in the Sheaves: The Harvest Excursionists, 1890-1929," *Canadian Historical Review*, LIX, 4 (December, 1978).

Table 3

Homesteads

A. Entries

	Man.	Sask.	Alta.	Prairies
1872	—	—	—	283
1873	—	—	—	878
1874	—	—	—	1,376
1875	—	—	—	499
1876	—	—	—	347
1877	—	—	—	845
1878	—	—	—	1,788
1879	—	—	—	4,068
1880	—	—	—	2,074
1881	—	23	—	2,753
1882	—	1,121	—	7,383
1883	—	—	—	6,063
1884	—	—	—	3,753
1885	—	—	—	1,858
1886	—	—	—	2,657
1887	1,053	356	271	1,680
1888	1,665	425	230	2,320
1889	2,225	1,242	504	3,971
1890	1,401	758	524	2,683
1891	1,651	930	784	3,365
1892	1,687	1,797	1,257	4,741
1893	1,276	1,159	1,513	3,948
1894	—	—	—	3,209
1895	866	461	1,000	2,327
1896	993	362	411	1,766
1897	609	301	230	1,140
1898	1,426	960	1,049	3,435
1899	2,124	2,159	1,745	6,028
1900	2,154	2,703	2,470	7,327
1901	1,933	2,332	3,806	8,071
1902	2,263	6,612	5,681	14,556
1903	3,253	19,941	8,069	31,263
1904	2,005	15,659	8,201	25,865
1905	1,707	19,787	9,138	30,632
1906	1,806	27,692	12,263	41,761

Table 3 – cont.

	Man.	Sask.	Alta.	Prairies
1907	1,231	13,501	6,843	21,575
1908	1,748	18,825	9,614	30,187
1909	3,761	21,120	13,771	38,652
1910	2,529	21,575	17,187	41,291
1911	3,082	25,227	15,964	44,273
1912	3,158	20,484	15,184	38,826
1913	2,826	17,556	12,942	33,324
1914	3,186	14,504	12,208	29,898
1915	4,420	8,790	10,076	23,286
1916	3,960	6,247	6,410	16,617
1917	2,276	4,105	4,550	10,931
1918	1,593	2,741	3,808	8,142
1919	813	1,191	2,169	4,173
1920	1,232	1,918	3,448	6,595
1921	725	1,670	2,874	5,269
1922	1,488	2,733	2,928	7,149
1923	879	2,104	2,207	5,190
1924	632	1,699	1,326	3,657
1925	464	1,804	1,192	3,460
1926	616	2,363	1,556	4,535
1927	797	2,702	2,145	5,644
1928	688	2,961	3,411	7,060
1929	643	5,808	8,933	15,384
1930	727	6,089	9,795	16,611
1931	454	2,834	7,122	10,410

B. Cancellations

	Man.	Sask.	Alta.	Prairies
1884	—	—	—	1,334
1885	—	—	—	—
1886	—	—	—	1,033
1887	—	—	—	633
1888	—	—	—	935
1889	—	—	—	1,337
1894	—	—	—	1,558

Table 3 – cont.

	Man.	Sask.	Alta.	Prairies
1895	—	—	—	1,222
1896	—	—	—	1,165
1897	—	—	—	1,090
1898	—	—	—	1,546
1899	—	—	—	1,746
1900	—	—	—	1,096
1901	—	—	—	1,682
1902	—	—	—	3,296
1903	—	—	—	5,208
1904	—	—	—	8,702
1905	—	—	—	11,296
1906	—	—	—	11,637
1907	—	—	—	14,110
1908	—	—	—	15,668
1909	—	—	—	14,677
1910	—	—	—	16,832
1911	—	—	—	22,122
1912	—	—	—	18,608
1913	2,006	8,288	6,694	16,988
1914	1,370	7,662	6,615	15,647
1915	1,694	4,953	5,432	12,079
1916	1,593	5,722	5,149	12,464
1917	1,578	3,558	4,101	9,237
1918	1,128	2,193	2,813	6,134
1919	929	1,100	1,946	3,975
1920	1,663	2,389	3,673	7,725
1921	1,403	2,360	3,365	7,128
1922	1,846	2,370	3,383	7,599
1923	1,895	2,278	2,652	6,825
1924	759	1,499	1,722	3,980
1925	1,025	1,475	1,552	4,052
1926	746	1,312	1,211	3,269
1927	547	1,330	1,489	3,366
1928	1,034	1,522	2,203	4,759
1929	654	1,717	2,741	5,112
1930	552	2,261	3,823	6,636
1931	381	1,219	2,428	4,028

SOURCE: M.C. Urquhart and K.A.H. Buckley, eds., *Historical Statistics of Canada* (Toronto, 1965), K34-41, p. 320; author's calculations. (The Manitoba entry figures for 1881 and 1882 appear unlikely and are not included here.)

Table 4

Prairie Farms

	No. of farms	Total acres (000s)	Total improved acres (000s)	Acres in field crops (000s)
A. Manitoba				
1881	9,077	2,384	250	230
1891	22,571	5,228	1,232	1,229
1901	32,252	8,843	3,995	2,756
1906	36,141	—	4,228	4,220
1911	43,631	12,184	6,746	5,162
1916	46,580	13,437	7,188	5,117
1921	53,252	14,616	8,057	5,858
1926	53,251	14,412	8,346	6,261
1931	54,199	15,132	8,522	5,842
B. Saskatchewan (incl. Alberta before 1901)				
1881	1,014	314	29	21
1891	9,244	2,910	197	190
1901	13,445	3,833	1,123	656
1906	55,971	—	—	3,271
1908	65,945	17,066	8,005	6,921
1911	95,013	28,099	11,872	9,137
1916	104,006	36,801	19,632	13,973
1921	119,451	44,023	25,037	17,822
1926	117,781	45,945	27,714	19,559
1931	136,472	55,673	33,549	22,126
C. Alberta (incl. in Saskatchewan before 1901)				
1901	9,479	2,736	475	188
1906	30,286	—	—	916
1911	60,559	17,359	4,352	3,378
1916	67,977	23,063	7,510	5,506
1921	82,954	29,293	11,768	8,523
1926	77,130	28,573	13,204	9,167
1931	97,408	38,977	17,749	12,037

Table 4 – cont.

	No. of farms	Total acres (000s)	Total improved acres (000s)	Acres in field crops (000s)
D. Prairies				
1881	10,091	2,698	279	251
1891	31,815	8,138	1,429	1,419
1901	55,176	15,412	5,593	3,600
1906	122,398	—	—	8,408
1911	199,203	57,643	22,970	17,677
1916	218,563	73,300	34,330	24,596
1921	255,657	87,932	44,863	32,203
1926	248,162	88,930	49,265	34,987
1931	288,079	119,783	59,819	40,066

SOURCES: *Census of Canada*, 1931, Vol. VIII, Table 1, p. xxviii; Manitoba Agriculture, *100 Years of Agriculture – Manitoba: A Statistical Profile, 1881-1980* (Winnipeg, 1981), pp. 10-11; Dominion Bureau of Statistics, *The Prairie Provinces in Their Relation to the National Economy of Canada: A Statistical Study of Their Social and Economic Condition in the Twentieth Century* (Ottawa, 1934), pp. 30-31; Saskatchewan Agriculture Report, 1909, p. 76.

Table 5

Mechanization

A. Tractor Sales

	Man.	Sask.	Alta.	Prairies
1919	3,627	3,514	1,703	8,844
1920	3,671	4,229	2,379	12,279
1921	1,057	1,655	716	3,428
1922	1,361	2,475	386	4,222
1923	911	2,524	731	4,166
1924	465	1,213	434	2,112
1925	1,008	2,176	869	4,053
1926	1,498	3,704	1,311	6,513
1927	1,414	5,727	2,885	10,026
1928	2,209	8,703	6,231	17,143
1929	2,423	6,906	5,228	14,557
1930	1,541	4,350	3,100	8,991
1931	186	267	334	787

B. Combine Sales

	Man.	Sask.	Alta.	Prairies
1926	2	148	26	176
1927	21	382	195	598
1928	206	2,356	1,095	3,657
1929	158	2,484	858	3,500
1930	134	939	541	1,614
1931	35	92	54	179

Table 5 – cont.

C. Numbers of Horses, Alberta (000s)

1906	226.5	1919	763.5
1907	255.0	1920	770.6
1908	280.9	1921	806.2
1909	314.5	1922	785.4
1910	365.1	1923	772.7
1911	407.2	1924	787.5
1912	429.5	1925	783.9
1913	483.2	1926	784.3
1914	539.0	1927	773.0
1915	585.6	1928	761.0
1916	629.5	1929	764.8
1917	660.2	1930	740.0
1918	720.5	1931	731.7

SOURCES: Canadian Farm Implements, December, 1931, cited in Andrew Stewart, "Trends in Farm Power and Their Influence on Agricultural Development," Appendix A in R.W. Murchie, *Agricultural Progress on the Prairie Frontier* (Toronto, 1936), pp. 296, 298. Alberta Dept. of Agriculture, *A Historical Series of Agricultural Statistics*, p. 58.

Notes

Abbreviations

GAI	Archives of the Glenbow-Alberta Institute (Calgary)
MLL	Manitoba Legislative Library (Winnipeg)
NAC	National Archives of Canada (Ottawa)
PAA	Provincial Archives of Alberta (Edmonton)
PAM	Provincial Archives of Manitoba (Winnipeg)
SAB	Saskatchewan Archives Board (Saskatoon and Regina)
TFL	Thomas Fisher Library (University of Toronto)
WRA	Walter P. Reuther Archives of Labor and Urban Affairs (Detroit)

Note on Sources

Hired hands have left few accounts of their world, so their story must be extracted largely from records that are not their own. This study has pieced it together from diverse sources: government reports, private agency records, farm journals, manuscript collections, pioneer memoirs, and, where possible, letters and diaries of hired hands.

Useful government records are those from provincial and federal departments of Immigration, Agriculture, and Labour and from offices of the Attorneys General. Records of businesses with interests in immigration and farming, such as the Canadian Pacific Railway, the Canadian National Railways, and various colonization companies, have yielded data on farm employment. Specific collections on labour, such as the papers of the Alberta Federation of Labour, the Industrial Workers of the World, and the Communist Party of Canada, have only occasionally turned up useful material, although this in itself is a telling comment on the place of farm workers in the labour movement, or rather, their absence from it.

While statistical compilations have been used wherever possible, they are in short supply and often of limited usefulness, as in the case of census records. Hard data have in places provided the bare bones of evidence, but these have been fleshed out by textual sources. Manuscript collections at the Provincial Archives of Manitoba, the Saskatchewan Archives Board, the Provincial Archives of Alberta, and the Archives of the Glenbow-Alberta Institute have been most helpful, containing a wealth of papers on rural life and the farming community. Farmers' journals and unpublished memoirs, farm account books, and records of farm organizations have provided much indirect information.

Seldom does the story come from the men themselves. Typical is the diary (SAB, RE-191) of prairie settler Geoffrey Yonge, who home-steaded near Mortlach, Saskatchewan. He carefully recorded the daily details of running his own farm, but when he departed every spring and fall to take up waged farm work, his diary fell silent. His entry on July 14, 1908, the first to appear after three months, tantalizes the researcher's imagination: "My time up with Cudmore, last night after a term of three months toil, from 4 AM to 9 PM almost every day. So I am glad to be free once more."

The rare diaries and letters of farm workers provide details unavailable in the quantitative sources. They are handled carefully, for they present experience through linguistic, cultural, and personal filters. Nonetheless, such documents provide insight into the lives of farm workers, enabling us to discover their hopes and ideas, to learn what they found serious or humorous, to explore as much as they choose to show us about themselves. Thus does the present study strive to reconstruct the lives of prairie hired hands.

Chapter 1

1. NAC, MG30, Vol. C133, George Becker Papers, unpublished ms. "Memories."
2. Canadian Pacific Railway, *Farming and Ranching in Western Canada* (Toronto, 1892), p. 25.
3. NAC, Becker Papers, unpublished ms. "Memories."
4. *Ibid.*, Vol. C165, Wilfred Bastyan Rowell Papers, Correspondence to his mother, 29/5/1904, 4/6/1904, 20/11/1904, 9/2/1905, 29/5/1905.
5. SAB, R249, Royal Commission on Immigration and Settlement (Saskatchewan), Records of Proceedings, Vol. VIII, Tisdale, testimony of Fred J. Watson, 20/2/1930, p. 47.

Chapter 2

1. SAB, M6/X-11-0, Charles Dunning Papers: Labour, p. 14562, "Farm Wages," March, 1922.

2. "Armstrong Report," in Greg Kealey, ed., *Canada Investigates Industrialism: The Royal Commission on the Relations of Labor and Capital, 1889 (Abridged)* (Toronto, 1973), p. 48.

3. Doug Owram, *Promise of Eden: The Canadian Expansionist Movement and the Idea of the West, 1856-1900* (Toronto, 1980), pp. 38-58; J.M.S. Careless, *Brown of the Globe: The Voice of Upper Canada, 1818-1859*, Vol. I (Toronto, 1959), p. 229.

4. R. Douglas Francis, *Images of the West: Responses to the Canadian Prairies* (Saskatoon, 1989), pp. 73-86.

5. Owram, *Promise of Eden*, pp. 59-78, esp. 60, 67-69, 76.

6. W.A. Waiser, *The Field Naturalist: John Macoun, the Geological Survey, and Natural Science* (Toronto, 1989), pp. 16-54.

7. Cited in Norman Macdonald, *Canada: Immigration and Colonization, 1841-1903* (Toronto, 1966), pp. 239-40.

8. Charles Horetzky, *Canada on the Pacific, being an Account of a Journey from Edmonton to the Pacific by the Peace River Valley* (Montreal, 1874), cited in Gordon E. Bowes, ed., *Peace River Chronicles* (Vancouver, 1963), p. 86.

9. For an explanation of the homestead system in western Canada and a comparison with the American system, see Chester Martin, *'Dominion Lands' Policy*, Lewis H. Thomas, ed. (Toronto, 1973), pp. 124-27, 140-42; originally published 1938.

10. See, for example, *ibid.*; William Marr and Michael Percy, "The Government and the Rate of Canadian Prairie Settlement," *Canadian Journal of Economics*, 11, 4 (1978); James M. Richtik, "The Policy Framework for Settling the Canadian West, 1870-1880," *Agricultural History*, 49, 4 (October, 1975); John Langton Tyman, *By Section, Township, and Range: Studies in Prairie Settlement* (Brandon, Manitoba, 1972).

11. See Appendix, Table 1, Agricultural Work Force.

12. The group under study is adult male farm workers. Women often appeared in the census in various agricultural pursuits such as general farming, dairying, and bee-keeping, but seldom as agricultural labourers. In the literature, they have appeared as farm hands so scarcely that their inclusion in this study would present a misleading picture of their experience. It warrants a separate study. The same is true for male children. They were sometimes listed as farm workers, although the literature suggests they usually did work classified as "chores." For purposes of this study, the references to hired hands will include only males aged fifteen and over, unless otherwise specified. Comparisons with other groups will be restricted to the same age and sex categories, unless otherwise indicated.

13. This difficulty of defining an agricultural labourer is reflected in the methods used to tally the census. After 1911, changes in the date of the census, from the first of March to the first of June, and changes in the definition of different types of agricultural labourers mean that categories do not correspond from one census year to the next. Moreover, the seasonal nature of agriculture makes these figures less than reliable whatever date or definition is used. As long as the census was taken at the first of March, agricultural labourers who found winter work elsewhere were excluded. A more accurate measurement was likely achieved when the census date was moved forward to the first of June, although short-term workers employed only for spring planting might have been included. The census definition of paid agricultural labourers changed slightly in different years. Even in a single census year, the designation of men as "agricultural labourers" or as "wage earners within the agricultural industry" results in different sets of figures. The latter group, but not the former, for example, included such employees as foremen on large farms. Another difficulty is that the census data include men who worked in agricultural enterprises such as labour-intensive sugar-beet farming, not just hired hands on mixed or grain farms. Wherever possible, the figures are adjusted for the purposes of this study to eliminate significant groups of men who do not fall into the more refined category of "permanent full-time hired hands." Finally, determining the actual numbers of agricultural labourers in the prairie provinces throughout the period of this study has presented special challenges. Census figures in 1881 and 1901 did not distinguish between paid farm workers and unpaid family members, and in 1891 did not specify their location within the North-West Territories. Despite these difficulties, however, the census data are useful for the trends they show.

14. John Herd Thompson and Allen Seager, "Workers, Growers and Monopolists: The 'Labour Problem' in the Alberta Sugar Beet Industry During the 1930s," *Labour/Le Travailleur*, 3 (1978).

15. John Herd Thompson, "'Bringing in the Sheaves': The Harvest Excursionists, 1890-1929," *Canadian Historical Review*, LIX, 4 (December, 1978).

Chapter 3

1. The words are those of prairie farmer G.A. Cameron of Indian Head, North-West Territories, as quoted in a CPR publication: [Alexander Begg], *Plain Facts from Farmers in the Canadian North-West* (London, 1885), p. 48.

2. PAM, MG 14/B30/2, Colin Inkster Papers, contract, 1874.

3. "The First Shipment of Grain From Manitoba," *Manitoba Daily Free Press*, 23/10/1876, p. 3.

4. Duncan A. MacGibbon, *The Canadian Grain Trade* (Toronto, 1932), p. 27.

5. W.L. Morton, "Agriculture in the Red River Colony," *Canadian Historical Review*, XXX, 4 (1949).

6. David Spector, *Agriculture on the Prairies, 1870-1940*, National Historic Parks and Sites Branch, Parks Canada, Department of the Environment (Ottawa, 1983); Ernest B. Ingles, "Some Aspects of Dry-Land Agriculture in the Canadian Prairies to 1925" (M.A. thesis, University of Calgary, 1973).

7. Wayne D. Rasmussen, "The Mechanization of Agriculture," *Scientific American*, 247, 3 (September, 1982).

8. O.L. Symes, "Chronology of Agricultural Machinery Development and Related Information," *Proceedings of Seminar: Development of Agriculture on the Prairies* (Regina, 1975), pp. 17, 20.

9. Archie A. Stone and Harold E. Gulvin, *Machines for Power Farming* (New York, 1957), p. 520.

10. Ernest B. Ingles, "The Custom Threshermen in Western Canada, 1890-1925," in David C. Jones and Ian MacPherson, eds., *Building Beyond the Homestead* (Calgary, 1985), p. 136.

11. See Appendix, Table 4, Prairie Farms.

12. *Canada Yearbook*, 1914, p. 181. For an explanation of crop statistics and a detailed series of tables, see Cecilia Danysk, "Against the Grain: Accommodation to Conflict in Labour-Capital Relations in Prairie Agriculture, 1880-1930" (Ph.D. thesis, McGill University, 1991), Appendix.

13. *Statistical Yearbook of Canada*, 1898, p. 70; Manitoba Department of Agriculture, *100 Years of Agriculture*, p. 18. Manitoba is used as an example here because its production figures are more reliable for the early period than are those of the North-West Territories.

14. Letter from a Dublin man [n.d.] Troy, NWT, cited in *The Emigrants' Guide for 1883* (London, 1883), p. 23.

15. Klaus Peter Stitch, "'Canada's Century': The Rhetoric of Propaganda," *Prairie Forum*, 1 (April, 1976); R. Douglas Francis, *Images of the West: Responses to the Canadian Prairies* (Saskatoon, 1989), pp. 107-55; Ronald A. Wells, ed., "Editor's Introduction," *Letters from a young emigrant in Manitoba* (Winnipeg, 1981), pp. 16-22; David Hall, *Clifford Sifton: The Young Napoleon, 1861-1900* (Vancouver, 1981), pp. 253-69. Immigration agents in Great Britain reported success in reaching prospective emigrants at "hiring markets, fairs, [and] agricultural shows." Robert Murdoch, "Report of the Glasgow Agency," 27/12/1875, *Report of the Minister of Agriculture for the Dominion of Canada for the Calendar Year 1875*, Dominion Sessional Paper 8, 1876, p. 82.

16. Cited in [Begg], *Plain Facts*, p. 48.

17. W.M. Champion, cited in [Alexander Begg], *Practical Hints from Farmers in the Canadian North-West* (London, 1885), p. 10.

18. See, for example *ibid*,. p. 9; Robert Christy Miller, *Manitoba Described: Being a Series of General Observations upon the Farming, Climate, Sport, Natural History, and Future Prospects of the Country* (London, 1885), p. 105; W. Henry Barneby, *Life and Labour in the Far, Far West: Being Notes of a Tour in the Western States, British Columbia, Manitoba, and the North-West Territory* (London, 1884), p. 338; Canada, Department of Agriculture and Immigration, *Manitoba. The Prairie Province. The finest agricultural country in the world* (Winnipeg, 1890); Canada, Department of the Interior, *Some of the Advantages of Western Canada. Practical Farmers give their experiences* (Ottawa, 1889); Canadian Pacific Railway, *Facts for Farmers. The great Canadian North-West; its climate, crops and capabilities; with settlers' letters* (Liverpool, 1887).

19. F. Woodcutter, "What Rev. F. Woodcutter, Parish Priest, has to say about his experiences in Canada, and especially about the Esterhaz colony, Kaposvar Post Office, Assiniboia, Canada, July, 1902" in [Paul Oscar Esterhazy], *The Hungarian Colony of Esterhaz, Assiniboia, North-west Territories, Canada* (Ottawa, 1902), reprinted in Martin Louis Kovacs, *Esterhazy and Early Hungarian Immigration to Canada: A Study Based Upon the Esterhazy Immigration Pamphlet* (Regina, 1974), p. 78.

20. Owram, *Promise of Eden*, p. 137.

21. [Begg], *Plain Facts*, p. 41.

22. R.G. Marchildon and Ian MacPherson, "'A Stout Heart and a Willing Mind': The Wherewithal of Prairie Settlers in the 1870s and 1880s," paper presented to the Canadian Historical Association annual meeting, Montreal, 1985, p. 13.

23. John Macoun, *Manitoba and the Great North-West: The Field for Investment. The Home of the Emigrant* (London, 1883), pp. 196, 146-47.

24. [Begg], *Plain Facts*, p. 3.

25. Macoun, *Manitoba and the Great North-West*, p. 642.

26. As well as *Plain Facts* and *Practical Hints*, already cited, other CPR titles were: *Manitoba – The Canadian North-west; the resources and capabilities of the Canadian North West as well as some experiences of men and women settlers* (Montreal, 1883); *Manitoba, the Canadian North-west: what the actual settlers say* (Montreal, 1886); *What Women Say of the Canadian North West* (Montreal, 1886).

27. For a detailed appraisal of the questionnaires, located in the Alexander Begg Papers at the Provincial Archives of British Columbia, see Marchildon and MacPherson, "'A Stout Heart and a Willing Mind.'" They argue that the questionnaires are an accurate representation of the experiences of British and British-Canadian settlers in the most populous regions of the North-West (pp. 6-7) and note that Begg did not publish unfavourable responses (p. 5).

28. [Begg], *Plain Facts*, pp. 5-6.

29. Barneby, *Life and Labour in the Far, Far West*, pp. 252-53. Barneby

calculated the value of a pound at $5; the actual value in 1884 was $4.73.

30. CPR, *The North-West Farmer in Manitoba, Assiniboia, Alberta* (1891), p. 10.

31. Reverend Nestor Dmytriw, "Canadian Ruthenia," in Harry Piniuta, ed. and trans., *Land of Pain, Land of Promise: First Person Accounts by Ukrainian Pioneers, 1891-1914* (Saskatoon, 1981), p. 47.

32. Robert E. Ankli and Robert Litt, "The Growth of Prairie Agriculture: Economic Considerations," *Canadian Papers in Rural History*, Vol. I, Donald H. Akenson, ed. (Gananoque, Ontario, 1980), p. 56; Lyle Dick, "Estimates of Farm-Making Costs in Saskatchewan, 1882-1914," *Prairie Forum*, 6, 2 (1981), p. 198.

33. Irene M. Spry, "The Cost of Making a Farm on the Prairies," *Prairie Forum*, 7, 1 (1982), pp. 95-96; and see Dick's reply in which he concedes the point: Lyle Dick, "A Reply to Professor Spry's Critique 'The Cost of Making a Farm on the Prairies,'" *Prairie Forum*, 7, 1 (1982), p. 101.

34. Alfred Pegler, *A Visit to Canada and the United States in Connection with the Meetings of the British Association Held in Montreal, in 1884* (Southampton, 1884), p. 26.

35. GAI, CH .C212B, CPR Papers, Advertisement for the North-West Territories, c. 1883.

36. Pegler, *A Visit to Canada*, p. 22.

37. *Ibid.*, pp. 16, 23.

38. Charles Foy, "Annual Report of the Belfast Agency 1875," *Report of the Minister of Agriculture . . . for 1875*, p. 83.

39. [Begg], *Plain Facts*, pp. 5-6.

40. Szikora Mihaly, "What Some of the Hungarian Settlers have to say for themselves, about the Canadian North-west, particularly about the Esterhaz colony, – their present abiding place, and of the productiveness of their own farms, and the cause of their prosperous circumstances in these days," in [Esterhazy], *The Hungarian Colony of Esterhaz*, p. 127.

41. W.E. Cooley, "Settlers' Testimony," Canada, Department of the Interior, *Western Canada: How to Get There; How to Select Lands; How to Make a Home* (1902).

42. Pegler, *A Visit to Canada*, p. 26.

43. *Report of the Commissioner of the North-West Mounted Police Force 1886*, reprinted in Royal North-West Mounted Police, *Law and Order 1886-1887* (Toronto, 1973), p. 17.

44. PAM, RG1/D1/VI, Department of Agriculture and Immigration Records, p. 361, J.S. Armitage to S.E. Brown, 11/7/1887.

45. *Ibid.*, p. 354, Armitage to Emerson Lorne, 5/7/1887.

46. James Kelly to A.J. McMillan, 23/8/1891?, cited in CPR, *Farming and Ranching in Western Canada. Manitoba, Assiniboia, Alberta, Saskatchewan* (1892?), p. 25.

47. J.G. Colmer, *Across the Canadian Prairies: A Two Months' Holiday in the Dominion* (London, 1895), p. 84.

48. Letter from Col. P.G.B. Lake and R.L. Lake, 20/10/1890, cited in CPR, *The North-West Farmer in Manitoba, Assiniboia, Alberta* (1891), p. 47.

49. Canada, Department of the Interior, *Western Canada. Manitoba and the Northwest Territories, Assiniboia, Alberta, Saskatchewan. Information as to the resources and climates of these countries for intending farmers, ranchers, etc.* (1899), p. 38.

50. Alfred J. Church, ed., *Making a Start in Canada: Letters From Two Young Emigrants* (London, 1889), p. 170.

51. Maldwyn A. Jones, "The Background to Emigration From Great Britain in the Nineteenth Century," *Perspectives in American History*, VII, (1973).

52. Peter A. Russell, "Upper Canada: A Poor Man's Country? Some Statistical Evidence," in Donald Akenson, ed., *Canadian Papers in Rural History*, Vol. III (Gananoque, Ontario, 1982).

53. David Gagan, *Hopeful Travellers: Families, Land and Social Change in Mid-Victorian Peel County, Canada West* (Toronto, 1981), p. 40.

54. The economic and social consequences were felt in Ontario. The value of land prices fell between 1884 and 1894 from $625,478,706 to $587,246,117, a decline attributed to the development of the West. Canada, *Statistical Yearbook of Canada*, 1 (1895), p. 303. I am indebted to Bob Beal for bringing this information to my attention.

   Young Ontario women began to lament the shortage of marriageable partners. Their refrain became a popular song complaining of "that plaguey pest, . . . the Great North-West," which drew away eligible men:

   > One by one they all clear out
   > Thinking to better themselves, no doubt,
   > Caring little how far they go
   > From the poor little girls of Ontario.

   Cited in Joseph Schull, *Ontario Since 1867* (Toronto, 1978), p. 133.

55. [Begg], *Plain Facts*, p. 42.

56. *Ibid.*, p. 48.

57. J.P.D. Dunbabin, "The 'Revolt of the Field': The Agricultural Labourers' Movement in the 1870s," *Past and Present*, XVI (November, 1963).

58. Pamela Horn, "Agricultural Trade Unionism and Emigration, 1872-1881," *Historical Journal*, XV, 1 (1972).

59. Timothy L. Demetrioff, "Joseph Arch and the Migration of English Agricultural Labourers to Ontario During the 1870s," unpublished paper, Queen's University, September, 1982, pp. 11-13.

60. *Ottawa Times*, 24/9/1873, cited *ibid.*, p. 10.

61. See *Report of the Minister of Agriculture . . . for 1875*, *passim*. The quotation is from Angus G. Nicholson, "Report of the Special Immigration Agent," Stornoway, 14/2/1876, p. 186.

62. *Ibid.*, A. Spencer Jones, "English Labourers' Union Delegate's Report," 14/1/1876, p. 106.

63. *Ibid.*, L. Letellier, p. xiv.

64. Edward Amey, *Farm Life As It Should Be and Farm Labourers' and Servant Girls' Grievances and Rules of the Proposed Agricultural Labourers' Union* (Toronto, 1885?), p. 25.

65. Alison Prentice, *The School Promoters: Education and Social Class in Mid-Nineteenth Century Upper Canada* (Toronto, 1977), p. 106.

66. *Ibid.* See also Tom Nesmith, "The Philosophy of Agriculture: The Promise of the Intellect in Ontario Farming, 1835-1925" (Ph.D. thesis, Carleton University, 1988).

67. SAB, R-418, George S. Tuxford Papers, Tuxford to Father and Mother, 29/9/1889.

68. Jones, "The Background to Emigration From Great Britain," p. 90.

69. [Begg], *Plain Facts*, p. 41.

70. George Thomas Haigh, "Report of the Liverpool Agent," 5/1/1876, *Report of the Minister of Agriculture . . . for 1875*, p. 75.

71. Jones, "English Labourers' Union Delegate's Report."

72. Karel Bicha, *The American Farmer and the Canadian West, 1896-1914* (Lawrence, Kansas, 1968), p. 16.

73. LaWanda F. Cox, "Tenancy in the United States, 1865-1900, A Consideration of the Validity of the Agricultural Ladder Hypothesis," *Agricultural History*, 3 (July, 1944), p. 102.

74. Margaret Beattie Bogue, *Patterns from the Sod: Land Use and Tenure in the Grand Prairie, 1850-1900* (Springfield, Illinois, 1959), pp. 161-75.

75. Peter H. Argersinger and Jo Ann E. Argersinger, "The Machine Breakers: Farmworkers and Social Change in the Rural Midwest of the 1870s," *Agricultural History*, 58, 3 (July, 1984), p. 396; Allen G. Applen, "Labor Casualization in Great Plains Wheat Production: 1865-1902," *Journal of the West*, XVI, 1 (January, 1977), pp. 5, 9.

76. Bogue, *Patterns from the Sod*, pp. 161-75, esp. 164.

77. Gilbert C. Fite, *The Farmers' Last Frontier, 1865-1900* (New York, 1966), p. 215.

78. See, for example, John-Paul Himka, "The Background to Emigration: Ukrainians of Galicia and Bukovyna, 1848-1914," in Manoly R. Lupul, ed., *A Heritage in Transition: Essays in the History of Ukrainians in Canada* (Toronto, 1982).

79. GAI, M364/1, Lars Peter Erickson Papers.

80. *Census of Canada*, 1880-81, Vol. I, Table IV, pp. 398-99; 1891, Vol. I, Table V, pp. 362-63; 1901, Vol. I, Table XIII, pp. 416-17.

81. Jones, "English Labourers' Union Delegate's Report."

82. *The Last Best West* was the title of an oft-reprinted Department of the Interior immigration pamphlet.

83. Lyle Dick, "Factors Affecting Prairie Settlement: A Case Study of

Abernethy, Saskatchewan in the 1880s," *Historical Papers/Communications historiques* (1985), p. 26.

84. Allen Smith, "The Myth of the Self-made Man in English Canada, 1850-1914," *Canadian Historical Review*, LIX, 2 (June, 1978), p. 192.

Chapter 4

1. GAI, A .5874/1, John Stokoe Papers, Stokoe to mother, 19/4/1903.

2. Harold Baldwin, *A Farm for Two Pounds: Being the Odyssey of an Emigrant* (London, 1935).

3. *Ibid.*, pp. 2, 18, 19, 21, 50, 55, 63, 74, 92, 93, 97, 120, 124, 258.

4. See Appendix, Table 4, Prairie Farms.

5. Manitoba Department of Agriculture, *100 Years of Agriculture*, p. 18; *Canada Yearbook*, 1905, p. 80; *Canada Yearbook*, 1916-17, pp. 192-93.

6. See Appendix, Table 3-A, Homestead Entries.

7. See Appendix, Table 1, Agricultural Work Force.

8. *Census of Canada*, 1931, Vol. I, Table 7a, p. 384.

9. Seager Wheeler, *Seager Wheeler's Book on Profitable Grain Growing* (Winnipeg, 1919); John Bracken, *Crop Production in Western Canada* (Winnipeg, 1920); John Bracken, *Dry Farming in Western Canada* (Winnipeg, 1921). For an assessment of the influence of Bracken and Wheeler on dry farming practices, see Spector, *Agriculture on the Prairies, 1870-1940*; Paul Voisey, *Vulcan: The Making of a Prairie Community* (Toronto, 1988), pp. 98-127.

10. David C. Jones, *Empire of Dust: Settling and Abandoning the Prairie Dry Belt* (Edmonton, 1987).

11. Royal Grain Inquiry Commission, *Report*, p. 23; *Canada Yearbook*, 1915, p. 200.

12. John Herd Thompson, *Harvests of War: The Prairie West, 1914-1918* (Toronto, 1978), pp. 47-48.

13. Royal Grain Inquiry Commission, *Report*, p. 23.

14. *Canada Yearbook*, 1915, pp. 161, 163, 165; *ibid.*, 1916-17, pp. 192-93.

15. The major crops are wheat, oats, barley, rye, mixed grains, and flaxseed. Dominion Bureau of Statistics, *Field Crops* (1975), pp. 28, 36, 40, 48, 52, 61, 76-77, 110, 114. Author's calculations. Farm value is the price paid to the farmer.

16. See Appendix, Table 4, Prairie Farms.

17. *Canada Yearbook*, 1915, pp. 161, 163, 165; *ibid.*, 1916-17, pp. 192-93.

18. Thompson, "'Bringing in the Sheaves.'"

19. Vernon C. Fowke, *The National Policy and the Wheat Economy* (Toronto, 1978), p. 74; originally published 1957.

20. Royal Grain Inquiry Commission, *Report*.

21. Ian MacPherson, *Each for All, A History of the Co-operative Movement in English Canada, 1900-1945* (Toronto, 1979).

22. United Farmers of Alberta, *Annual Report and Year Book*, 1916, pp. 13, 17.

23. Lyle Dick, *Farmers 'Making Good': The Development of Abernethy District, Saskatchewan, 1880-1920*, Studies in Archaeology, Architecture and History: National Historic Parks and Sites, Canadian Parks Service, Environment Canada, 1989, pp. 123-30; Ian MacPherson and John Herd Thompson, "The Business of Agriculture: Farmers and the Adoption of 'Business Methods', 1880-1950," in Peter Baskerville, ed., *Canadian Papers in Business History*, Vol. I (Victoria, 1989), p. 247.

24. Bogue, *Patterns from the Sod*, pp. 161-69.

25. William L. Marr, "Tenant vs. Owner Occupied Farms in York County, Ontario, 1871," in Donald Akenson, ed., *Canadian Papers in Rural History*, Vol. IV (Gananoque, Ontario, 1984); Joy Parr, "Hired Men: Ontario Agricultural Wage Labour in Comparative Perspective," *Labour/Le Travail*, 15 (1985), p. 95.

26. United States Bureau of the Census, *Twelfth Census of the United States, Taken in the Year 1900: Statistical Atlas* (Washington, 1903), pp. 72-73, plate 140.

27. Frederick Jackson Turner, *The Significance of the Frontier in American History*, ed. and intro. by Harold P. Simonson (New York, 1976), pp. 51, 52, 57; originally published 1893.

28. Bicha, *The American Farmer and the Canadian West*, p. 16.

29. GAI, M1008/A .P957, Frederick Pringle Papers, diary, p. 111.

30. *Ibid.*, pp. 126, 128.

31. *Ibid.*, pp. 129, 140-41.

32. R.W. Murchie, *Agricultural Progress on the Prairie Frontier* (Toronto, 1936), pp. 195, 262, 242, 156.

33. *Ibid.*, pp. 242, 259, 289, 156, 159. Author's calculations.

34. See, for example, SAB, RE 191, Geoffrey Yonge Papers; GAI, A.B474A, Roy Benson Papers.

35. Even in the late 1920s, homesteaders in newly settled areas were found to "derive their income from off-farm wages," compared with farmers in more settled areas who earned most of their income from "the farms on which they live." C.A. Dawson and R.W. Murchie, *The Settlement of the Peace River Country: A Study of a Pioneer Area* (Toronto, 1934), pp. 112-13; C.A. Dawson and Eva Younge, *Pioneering in the Prairie Provinces: The Social Side of the Settlement Process* (Toronto, 1940), p. 138.

36. Murchie, *Agricultural Progress on the Prairie Frontier*, pp. 156, 242.

37. Percy Maxwell to Eveline and Stanley, 4/4/1904, in Percy Augustus Maxwell, *Letters home during his years as a homesteader in the developing period of Canada's West* (printed for private circulation, 1967), p. 65.

38. "Labour Laws of Manitoba," The Masters and Servants Act, Chapter 124

with amendments, in Canada, Department of Labour, *Labour Legislation Existing in Canada in 1920* (Ottawa, 1921), pp. 471-73.

39. *Census of Canada*, 1911, Vol. V, Table V, pp. 52-53, 96-97, 228-29. Author's calculations.

40. Harry Braverman reckoned that apprenticeship for farming extends well beyond the three to seven years required for traditional crafts, covering "most of childhood, adolescence, and young adulthood." Harry Braverman, *Labor and Monopoly Capital: The Degradation of Work in the Twentieth Century* (New York, 1974), p. 109. Contemporary observers and immigration and settlement agents were more sanguine, seldom suggesting more than a year or two.

41. *Labour Gazette*, 1901, p. 560.

42. See Appendix, Table 2-A, Annual Farm Workers' Wages.

43. *Labour Gazette*, June, 1901, p. 561.

44. Dick, *Farmers 'Making Good'*, p. 66.

45. SAB, X2, Pioneer Questionnaires, #6 Pioneer Farming Experiences, author's calculations. Beginning in the winter of 1950-51, the Saskatchewan Archives Board distributed to early settlers a series of questionnaires dealing with pioneer life. The number of responses varied with each questionnaire from approximately 300 for #8 Health to approximately 900 for #2 General.

46. SAB, M1/IV Walter Scott Papers: Labour, p. 46050, "Unemployment."

47. Willem de Gelder, *A Dutch Homesteader on the Prairies: The Letters of Willem de Gelder, 1910-13*, trans. and intro. by Herman Ganzevoort (Toronto, 1973), p. 5. See also Wells, *Letters from a young emigrant*, pp. 72-77; Shepherd, *West of Yesterday*, p. 18; GAI, A .5874/1, John Stokoe Papers; SAB, RE 33, Harry Self Papers, "Memories of my First Years in Canada," 1964.

48. de Gelder, *A Dutch Homesteader on the Prairies*, p. 46.

49. SAB, Harry Self Papers, "Memories of my First Years in Canada," pp. 2, 5, 7.

50. Gaston Giscard, *Dans La Prairie Canadienne*, trans. by Lloyd Person, ed. by George E. Durocher (Regina, 1982), p. 14.

51. GAI, John Stokoe Papers, Stokoe to mother, 19/4/1903.

52. *Ibid.*

53. *Ibid.*, Stokoe to Adam, 28/10/1905; Stokoe to father, 6/6/1906; Stokoe to father, 11/9/1904.

54. *Ibid.*, Stokoe to father, 23/10/1906.

Chapter 5

1. NAC, MG31, H15, S. Jickling, "Hoosier Valley," ms., p. 93.

2. SAB, Pamphlet File, Biography, Charles Fisher, "The Accepted Immigrant," Section 2, "Decision to Homestead," ms., p. 3.

3. *Ibid.*, pp. 3-5; Section 3, "The Hired Man," pp. 14-23, 29.

4. E.P. Thompson, *The Making of the English Working Class* (New York, 1982), originally published 1963; Herbert Gutman, ed., *Work, Culture and Society in Industrializing America, 1815-1919: Essays in American Working-Class and Social History* (New York, 1976); Greg Kealey, *Toronto Workers Respond to Industrial Capitalism, 1867-1892* (Toronto, 1980); Bryan Palmer, *A Culture in Conflict: Skilled Workers and Industrial Capitalism in Hamilton, Ontario, 1860-1914* (Montreal, 1979). For an attack on that position and the methodology, see David J. Bercuson, "Through the Looking Glass of Culture: An Essay on the New Labour History and Working-Class Culture in Recent Canadian Historical Writing," *Labour/Le Travailleur*, 7 (Spring, 1981).

5. See, for example, Leo A. Johnson, "Independent Commodity Production: Mode of Production or Capitalist Class Formation?" *Studies in Political Economy*, 6 (Autumn, 1981); J. Chevalier, "There is Nothing Simple About Simple Commodity Production," *Studies in Political Economy*, 7 (Winter, 1982); W. Denis, "Capital and Agriculture: A Review of Marxian Problematics," *Studies in Political Economy*, 7 (Winter, 1982).

6. Cecilia Danysk, "Farm Apprentice to Agricultural Proletarian: The Hired Hand in Alberta, 1880-1930" (M.A. thesis, McGill University, 1981); Gerald Friesen, *The Canadian Prairies: A History* (Toronto, 1984), pp. 316-20; Voisey, *Vulcan*, pp. 221-24; Dick, *Farmers 'Making Good'*, pp. 123-25, quotation from p. 124; W.J.C. Cherwinski "In Search of Jake Trumper: The Farm Hand and the Prairie Farm Family," in David C. Jones and Ian MacPherson, eds., *Building Beyond the Homestead* (Calgary, 1985); Thompson and Seager, "Workers, Growers and Monopolists." Compare with the view of proletarian behaviour in Argersinger and Argersinger, "The Machine Breakers"; Allen Seager, "Captain Swing in Ontario?" *Bulletin of the Committee on Canadian Labour History*, 7 (Spring, 1979). I am currently engaged in a detailed study of agricultural labour on giant "bonanza farms" in the prairie West, 1880s to 1930s, which will shed light on the situation of farm workers in a setting akin to factory production.

7. Raymond Williams, *Keywords: A Vocabulary of Culture and Society* (New York, 1976), p. 129.

8. Thompson, *Making of the English Working Class*, p. 9.

9. John Herd Thompson and Ian MacPherson, "How you Gonna Get 'em Back to the Farm?: Writing the Rural/Agricultural History of the Prairie West," paper presented to the Western Canadian Studies Conference, Saskatoon, 1987, p. 11.

10. Abundant examples are provided in local histories and pioneer reminiscences, such as A. Bert Reynolds, *'Siding 16': An Early History of Wetaskiwin to 1930* (Wetaskiwin, Alberta, 1975), p. 220; Weyburn R.M. 67 History Book Committee, *As Far as the Eye Can See: Weyburn R.M.*

67 (Weyburn, Sask., 1986); James M. Minifie, *Homesteader: A Prairie Boyhood Recalled* (Toronto, 1972).

11.   GAI, M 1345/A.W949, Frederick John Wright interview, February, 1971.

12.   Shepherd, *West of Yesterday*, p. 23.

13.   See, for example, SAB, B89, Agricultural Societies, 1912-1933; PAA, 73.316, Agricultural Societies, 1912-1933, esp. 4/26a,b, Hays Agricultural Society, 27a, High Prairie Agricultural Society, 30a, Lacombe Agricultural Society, 53a, Rocky Mountain House Agricultural Society, 58a, Stettler Agricultural Society.

14.   See, for example, PAA, 66.119, Norman Nelson Papers, diary of Wm. Sutton, hired hand; GAI, A .5874/1, John Stokoe Papers; PAA, 73.167, Jens Skinberg interview, 31/1/1973.

15.   See, for example, SAB, X2, Pioneer Questionnaires. Several articles in *Saskatchewan History* provide information gleaned from the questionnaires, such as Kathleen M. Taggart, "The First Shelter of Early Pioneers," XI, 3 (Autumn, 1958); Catherine Tulloch, "Pioneering Reading," XII, 3 (Autumn, 1959); Christine MacDonald, "Pioneer Church Life in Saskatchewan," XIII, 1 (Winter, 1960); E.C. Morgan, "Pioneer Recreation and Social Life," XVII, 2 (Spring, 1965).

16.   For a contrast of the positions, see the work on W.R. Motherwell: Sarah Carter "Material Culture and the W.R. Motherwell Home," *Prairie Forum*, 8, 1 (1983); Allan R. Turner, "W.R. Motherwell: The Emergence of a Farm Leader," *Saskatchewan History*, 11, 3 (Autumn, 1958); Dick, *Farmers 'Making Good'*. Compare with Voisey, *Vulcan*, pp. 201-46.

17.   D. McGinnis, "Farm Labour in Transition: Occupational Structure and Economic Dependency in Alberta, 1921-1951," in Howard Palmer, ed., *The Settlement of the West* (Calgary, 1977), p. 177.

18.   See, for example, SAB, X2, Pioneer Questionnaires, #5 Recreation and Social Life and #7 Folklore; local histories such as Barons History Book Club, *Wheat Heart of the West* (Barons, Alberta, 1972); La Riviere Historical Book Society (Lois Creith, convener), *Turning Leaves: A History of La Riviere and District* (La Riviere, Manitoba, 1979); and pioneer reminiscences such as GAI, D920.C554/1, Otto D. Christensen Papers; SAB, RE 191, Geoffrey Yonge Papers.

19.   Cited in John H. Blackburn, *Land of Promise*, ed. and intro. by John Archer (Toronto, 1970), p. 80.

20.   See, for example, Steven Penfold, "'Have You No Manhood in You?': Gender and Class in the Cape Breton Coal Towns, 1920-1926," *Labour/Le Travail*, 31 (Spring, 1993); Shirley Tillotson, "'We may all soon be first class men': Gender and Skill in Canada's Early Twentieth Century Urban Telegraph Industry," *Labour/Le Travail*, 27 (Spring, 1991); Cynthia Cockburn, *Brothers: Male Dominance and Technological Change* (London, 1983); Paul Willis, "Shop Floor Culture, Masculinity and the Wage Form," in John Clarke, Chas Critcher, and Richard

Johnson, eds., *Working Class Culture: Studies in History and Theory* (London, 1979); Joy Parr, *Gender of Breadwinners: Women, Men, and Change in Two Industrial Towns, 1880-1950* (Toronto, 1990); Mark Rosenfeld, "'It Was A Hard Life': Class and Gender in the Work and Family Rhythms of a Railway Town, 1920-1950s," *Historical Papers/Communications historiques* (Windsor, 1988); Steven Maynard, "Rough Work and Rugged Men: The Social Construction of Masculinity in Working-Class History," *Labour/Le Travail*, 23 (Spring, 1989).

21. Max Hedley, "Relations of Production of the 'Family Farm': Canadian Prairies," *Journal of Peasant Studies*, 9, 1 (October, 1981); Harriet Friedmann, "World Market, State, and Family Farm: Social Bases of Household Production in the Era of Wage Labor," *Comparative Studies in Sociology and History*, 20 (1978).

22. Doug Francis, *Images of the West: Responses to the Canadian Prairies* (Saskatoon, 1989), p. 233.

23. Elizabeth B. Mitchell, *In Western Canada Before the War: Impressions of Early Twentieth Century Prairie Communities* (Saskatoon, 1981), p. 150n; originally published 1915.

24. Eggleston, "The Old Homestead," p. 117.

25. Mrs. George [Marian] Cran, *A Woman in Canada* (Toronto, c. 1908).

26. Jean Burnet, *Next-Year Country: A Study of Rural Social Organization in Alberta* (Toronto, 1951), p. 21.

27. Edward ffolkes to mother, 15/12/1881, in *Letters from a young emigrant in Manitoba*, ed. and intro. by Ronald Wells (Winnipeg, 1981), p. 100.

28. *Ibid.*

29. "Bachelor Stock Raising," *Nor'-West Farmer*, 6/2/1899, p. 94.

30. *Hanna Herald*, 5/6/1913, cited in Burnet, *Next-Year Country*, p. 19.

31. Ebe Koeppen, diary, reprinted in Rolf Knight, *Stump Ranch Chronicles and Other Narratives* (Vancouver, 1977), pp. 120, 121, 129.

32. Leonora M. Pauls, "The English Language Folk and Traditional Songs of Alberta: Collection and Analysis for Teaching Purposes" (M.Mus., University of Calgary, 1981), p. 144.

33. Mitchell, *In Western Canada Before the War*, p. 47.

34. Koeppen in Knight, *Stump Ranch Chronicles*, p. 58.

35. Cited in Barry Broadfoot, *The Pioneer Years, 1895-1914: Memories of Settlers Who Opened the West* (Toronto, 1976), p. 138.

36. *Qu'Appelle Vidette*, 1/4/1885, p. 3.

37. In 1911, adult males (twenty-one years and over) outnumbered females by 137.7 to 100 in Manitoba, by 181.2 to 100 in Saskatchewan, and by 184.3 to 100 in Alberta. *Census of Canada*, 1921, Vol. II, Table 26, p. 124.

38. SAB, X2, Pioneer Questionnaires, #5 Recreation and Social Life.

39. Dawson and Younge, *Pioneering in the Prairie Provinces*, p. 310.

40. Felix Troughton, *A Bachelor's Paradise or Life on the Canadian Prairie 45 years ago* (London, c. 1930), p. 5.

41. *Ibid.*, pp. 5-6.

42. *Ibid.*, pp. 4, 6.

43. *Ibid.*, p. 6.

44. Mitchell, *In Western Canada Before the War*, p. 46.

45. Marjorie Harrison, *Go West – Go Wise! A Canadian Revelation* (London, 1930), p. 134.

46. ffolkes to mother, 15/12/1881, in Wells, *Letters from a young emigrant*, p. 100.

47. Koeppen in Knight, *Stump Ranch Chronicles*, p. 66.

48. Kathleen Strange, *With the West in Her Eyes: The Story of a Modern Pioneer* (Toronto, 1937), p. vii.

49. Baldwin, *A Farm for Two Pounds*, p. 45.

50. "Ruskin on Labor," *Nor'-West Farmer*, 6/2/1899, p. 100.

51. Compare, for example, with the response by male typographical workers to the threatened encroachment upon their trade by female workers. Cockburn, *Brothers*.

52. For a discussion of the muting of gender antagonisms through mutually supportive gender-distinguished work roles, see Ella Johansson, "Beautiful Men, Fine Women and Good Work People: Gender and Skill in Northern Sweden, 1850-1950," *Gender and History*, 1, 2 (Summer, 1989).

53. PAM, RG1/D1/VI, Department of Agriculture and Immigration Records, J.S. Armitage to Samuel Kennedy, 13/7/1887, p. 363.

54. Michael Ewanchuk, *Spruce, Swamp and Stone: A History of the Pioneer Ukrainian Settlements in the Gimli Area* (Winnipeg, 1977), p. 129.

55. Jaroslav Petryshyn, *Peasants in the Promised Land: Canada and the Ukrainians, 1891-1914* (Toronto, 1985), p. 115.

56. NAC, RG76/132/29490, Part 1, clipping from *Manitoba Free Press*, 30/7/1909.

57. Mabel F. Timlin, "Canada's Immigration Policy, 1896-1910," *Canadian Journal of Economics and Political Science*, XXVI, 4 (November, 1960), p. 517.

58. NAC, RG76/132/29490, Part 2, Scott to Davies, 21/7/1911.

59. *Ibid.*,1/8/1911.

60. *Ibid.*, Part 3, White to Gingras, 6/7/1912; Part 5, Scott to Perkins, 21/18/1911.

61. *Ibid.*, Part 3,White to Gingras, 6/7/1912.

62. Harold M. Troper, "The Creek-Negroes of Oklahoma and Canadian Immigration, 1909-11," *Canadian Historical Review*, LIII, 3 (September, 1972).

63. NAC, RG76/132/29490, Part 2, Scott to Othson, 27/7/1911.

64. *Ibid.*, Part 4, Strau to Scott, 29/6/1915.

65. PAA, 71.420/10, United Farmers of Alberta, "Resolution from Streamstown Local, No. 98," *Annual Report 1917*, p. 191.

66. See Chapter 6 for a fuller discussion.
67. PAA, 65.118/15/12, Redshaw to Smitten, 5/5/1927.
68. GAI, BN.C212G, Colley Papers, Farm Help Applications, author's calcu-
    lations. I am indebted to John Herd Thompson and Katrin Thompson for
    making available to me the material upon which the calculations in this
    section are based.
69. GAI, BN.C212G/1488, Pool to Colley, 13/6/1928.
70. See, for example, Howard Palmer, *Patterns of Prejudice: A History of
    Nativism in Alberta* (Toronto, 1982).
71. NAC, RG76/234/135755, Part 4, Memo, "Assisted Passage British Farm
    Workers," c. 1925.

Chapter 6

1. Some of the material in this chapter and in Chapter 7 first appeared in
   Cecilia Danysk, "'Showing These Slaves Their Class Position': Barriers
   to Organizing Prairie Farm Workers," in David C. Jones and Ian McPher-
   son, eds. *Building Beyond the Homestead* (Calgary, 1985). Permission of
   the University of Calgary Press to reprint this material is gratefully
   acknowledged. MLL, "Local History – Manitou," Acc: 8/1/1951, "A
   Citizen of Canada," ms. by W.N. Rolfe, p. 9.
2. GAI, M1008/A .P957, Frederick Pringle Papers, diary, p. 129.
3. *Ibid.*, pp. 130-31.
4. Reported in "Effects of Agricultural Depression on Farm Wages," *Labour
   Gazette*, June, 1931, p. 647. See also "Wages and Hours of Labour in
   Canada 1901-1920," Supplement to *Labour Gazette*, March, 1921, pp. 10-
   16; "Wages and Hours of Labour in Canada," Addendum on Wages of
   Coal Miners, 1900-1921, Supplement to *Labour Gazette*, February, 1922,
   p. 25.
5. Anna Farion, "Homestead Girl," in Harry Piniuta, ed. and trans., *Land
   of Pain, Land of Promise: First Person Accounts by Ukrainian Pioneers,
   1891-1914*, (Saskatoon, 1981), p. 87.
6. SAB, R418, George Tuxford Papers, p. 1092, Tuxford to father and mother,
   12/10/1911.
7. Koeppen in Knight, *Stump Ranch Chronicles,* p. 68.
8. Saskatchewan, Department of Agriculture, *Eighth Annual Report of the
   Department of Agriculture of the Province of Saskatchewan 1912* (Regina,
   1913), p. 137; *Ninth Annual Report of the Department of Agriculture of
   the Province of Saskatchewan 1913* (Regina, 1914), p. 195.
9. See Appendix, Table 2-A, Annual Farm Workers' Wages, and Figure 1.
10. *Tenth Annual Report of the Department of Agriculture of the Province of
    Saskatchewan 1914* (Regina, 1915), p. 137.
11. *Labour Gazette*, June, 1901, p. 556.
12. North-West Territories Department of Agriculture, *Annual Report of the*

Department of Agriculture of the North-West Territories, 1903 (Regina, 1904), p. 37.

13. *Labour Gazette*, June, 1901, p. 556.

14. Saskatchewan, Department of Agriculture, *Eighth Annual Report*, p. 137.

15. Giscard, *Dans La Prairie Canadienne*, p. 13.

16. Shepherd, *West of Yesterday*, p. 18.

17. *Labour Gazette*, June, 1901, pp. 555-56.

18. Saskatchewan, Department of Agriculture, Bureau of Labour, *Annual Report*, 1911, p. 20.

19. SAB, M4/I/100, W.M. Martin Papers: Labour, p. 29832, G. Graham to W.M. Martin, 18/3/1918.

20. See, for example, PAM, MG10/E1, United Farmers of Manitoba Papers, especially Minutes of Conventions; SAB, M12/II/60, W.R. Motherwell Papers.

21. Georgina Binnie-Clark, *Wheat and Woman*, intro. by Susan Jackel (Toronto, 1979; originally published 1914), p. 35.

22. Shepherd, *West of Yesterday*, p. 18.

23. *Labour Gazette*, 1901, pp. 564-65.

24. *Ibid.*, pp. 265-66.

25. See, for example, PAA, 65.118/17/47, Employment Service of Canada Papers, W.D. Trego, United Farmers of Alberta, Labor Committee, *Official Circular #10*, 29/3/1921; SAB, M6/X-11-0, Dunning Papers: Labour, p. 14563, "Farm Wages," March, 1922, and other submissions in the Annual Western Conferences.

26. Howard Newby, *The Deferential Worker: A Study of Farm Workers in East Anglia* (London, 1977), p. 279. Although farm workers in the prairie West demonstrated few of the deferential traits Newby found among East Anglian farm workers, prairie hands shared the perception that their work was infused with an "air of uniqueness."

27. Robert Collins, *Butter Down the Well: Recollections of a Canadian Childhood* (Saskatoon, 1980), p. 66.

28. The term "next-year country" was used so commonly by western farmers to describe their land that it has entered the prairie vocabulary. Jean Burnet's study of rural social organization in Alberta appropriately bears the phrase as its title.

29. Koeppen in Knight, *Stump Ranch Chronicles*, p. 49.

30. Despite the general belief that farm work was healthy, accidents were common. In July, 1904, for example, of the twenty-one accidents in agriculture reported in Canada, twelve were fatal. In contrast, the fatality rate was only five out of seventeen in mining, twelve out of thirty-six in railway service, five out of twenty-six in the building trades, and four out of twenty-five in unskilled labour. *Labour Gazette*, July, 1904, p. 98.

31. Murchie, *Agricultural Progress on the Prairie Frontier*, pp. 156, 262, 290.

32. ffolkes to mother, 6/10/1881, in Wells, *Letters from a young emigrant*, p. 83.

33. NAC, MG30 C63, Noel Copping Papers, "Prairie Wool and some Mosquitoes," excerpts from a diary, Saskatchewan, 1909-1910, pp. 18-19.

34. The material in this section is a composite of information gleaned from contemporary agricultural guides and farm journals, such as Bracken, *Crop Production in Western Canada*; Bracken, *Dry Farming in Western Canada*; Wheeler, *Seager Wheeler's Book on Profitable Grain Growing*; *Farm and Ranch Review*; *Nor'-West Farmer*; *Farmers' Advocate*; *Grain Growers' Guide*.

35. See Wheeler, *Profitable Grain Growing*, ch. III, esp, pp. 110-12, for the detailed care needed in cultivating. Not all land was backset.

36. PAM, MG8/B41, Arthur Jan Papers.

37. PAM, MG8/B18, John Cowell Papers, p. 2.

38. Weyburn R.M. 67 History Book Committee, *As Far as the Eye Can See*, p. 696.

39. S.J. Ferns and H.S. Ferns, *Eighty Five Years in Canada* (Winnipeg, 1978), p. 44.

40. SAB, C55, Philip Golumbia interview by D.H. Bocking, 23/12/1970.

41. Bernard Harmstone interview with author, Calgary, 8/10/1983, and "The Harvest Excursion. August 1926," ms. in author's possession, p. 1.

42. Ernest B. Ingles, "The Custom Threshermen in Western Canada," in David C. Jones and Ian MacPherson, eds., *Building Beyond the Homestead* (Calgary, 1985).

43. Ferns and Ferns, *Eighty Five Years in Canada*, p. 53.

44. Edward Corcoran, "My Experiences as a Farm Hand in Canada," *United Empire: The Journal of the Royal Empire Society*, XX (New Series) 3 (March, 1929), p. 151.

45. Maxwell to mother, 29/11/1903, in Maxwell, *Letters home during his years as a homesteader*, p. 44.

46. *Ibid.*, 18/1/1904, p. 53.

47. SAB, II/A/49, Ray Coates Reminiscences, ms., "To the Golden West 1903-1931," p. 10.

48. ffolkes to mother, 14/8/1881, in Wells, *Letters from a young emigrant*, p. 82.

49. Newell LeRoy Simms, *Elements of Rural Sociology* (New York, 1946), p. 429.

50. MLL, "Local History – Manitou," Acc: 8/1/1951, Rolfe, "A Citizen of Canada," p. 3. I am indebted to Donald Loveridge for bringing this manuscript to my attention.

51. James W. Rinehart, *The Tyranny of Work: Alienation and the Labour Process*, 2nd edition (Toronto, 1987); Braverman, *Labor and Monopoly Capital,* pp. 45-58. For a study of modern migrant farm workers, which finds they have a surprisingly low level of social alienation, see William

A. Rushing, *Class, Culture, and Alienation: A Study of Farmers and Farm Workers* (Lexington, Mass., 1972).

52. Maxwell to mother, 10/5/1903, in Maxwell, *Letters home during his years as a homesteader*, p. 11.

53. Shepherd, *West of Yesterday*, p. 18.

54. GAI, A.5874, John Stokoe Correspondence, Stokoe to father, 9/8/03.

55. Ferns and Ferns, *Eighty Five Years in Canada*, p. 53.

56. NAC, MG30 C63, Noel Copping Papers, "Prairie Wool and Some Mosquitoes," p. 18.

57. Giscard, *Dans La Prairie Canadienne*, p. 13.

58. GAI, M1008/A .P957, Frederick Pringle Papers, p. 144.

59. Koeppen in Knight, *Stump Ranch Chronicles*, p. 53.

60. Baldwin, *A Farm for Two Pounds*, p. 262.

61. Giscard, *Dans La Prairie Canadienne*, pp. 149-50.

62. Ian Radforth, *Bushworkers and Bosses: Logging in Northern Ontario, 1900-1980* (Toronto, 1987); Craig Heron, *Working in Steel: The Early Years in Canada, 1883-1935* (Toronto, 1988).

63. Binnie-Clark, *Wheat and Woman*, p. 247.

64. PAM, MG 8/B9, Charles Drury Papers, pp. 22, 24.

65. See Appendix, Table 3-A, Homestead Entries. This was lower than the rate for the previous decade. But the period 1904-13 was a time of rapid immigration when good lands were still plentiful.

66. "Homesteaders as Farm Hands," *Agricultural Gazette of Canada*, Vol. 4 (1917), p. 252.

67. "Farm Workmen," *Farmer's Advocate and Home Journal*, 8/12/1915, p. 1530.

68. See Appendix, Table 2, Farm Workers' Wages.

69. *Ibid.*

70. John Herd Thompson, "'Permanently Wasteful but Immediately Profitable': Prairie Agriculture and the Great War," *Historical Papers/Communications historiques*, 1976, p. 198.

71. PAM, MG 10/E1, United Farmers of Manitoba Papers, Box 5, "Resolutions to be considered at the Grain Growers' Convention, Brandon 9-11/1/1918."

72. SAB, 266/1/1609, Department of Agriculture, Statistics Branch, Movement of Threshing Machines, E. Oliver to Geo. Adams, 28/10/1915.

73. SAB, M1/IV, Walter Scott Papers: Labour, p. 46109, "Petition" from The Balcarres Grain Growers' Cooperative Association, 1916.

74. *Ibid.*

75. On Ukrainian workers, for example, see John Herd Thompson, "The Enemy Alien and the Canadian General Election of 1917," in John Herd Thompson and Frances Swyripa, eds., *Loyalties in Conflict: Ukrainians in Canada during the Great War* (Edmonton, 1983), pp. 29-30.

76. PAM, MG 13/HI, T.C. Norris Papers, p. 931, "Resolution" from

Municipality of North Cypress, 4/8/1917; see also PAM, MG 14/B45, Valentine Winkler Papers.

77. NAC, RG76, Immigration Branch, Vol. 132, File 29490, Part 7, "Extract from a weekly letter from Mr. Bruce Walker to Mr. Scott," 24/8/1917.

78. PAM, MG10/E1, United Farmers of Manitoba Papers, Box 15, "Report of Proceedings of Annual Convention," *Manitoba Grain Growers' Year Book, 1918*, p. 61.

79. PAM, MG 14/B45, Valentine Winkler Papers, p. 2379, Winkler to W.R. Wood, 11/7/1917.

80. Saskatchewan Bureau of Labour, *Annual Report*, 1916, p. 22.

81. *Agricultural Gazette of Canada*, Vol. 3 (1916), p. 248.

82. PAA, 84.407/65, W.J. Blair Papers, J.B. Burgeson to W.J. Blair, 25/7/1916.

83. Saskatchewan Bureau of Labour, *Annual Report*, 1916, p. 21.

84. *Agricultural Gazette of Canada*, Vol. 3, (1916), p. 248.

85. SAB, M4/I/101, W.M. Martin Papers: Labour: Farm: N. De Wind, "Patriotic Harvesting Clubs," pp. 30079-82. See also the offer from the Canadian Credit Men's Trust Association of its members' services to the Canadian Council of Agriculture. PAM, MG 10/E2/1, Canadian Council of Agriculture, *Minutes*, Winnipeg, 1/2/1917.

86. *The Albertan* [Calgary], 22/8/1917. For reasons for this fear, see David Schulze, "The Industrial Workers of the World and the Unemployed in Edmonton and Calgary in the Depression of 1913-1915," *Labour/Le Travail*, 25 (Spring, 1990).

87. Philip Taft, "The I.W.W. in the Grain Belt," in Stanley Coben and Forest G. Hill, eds., *American Economic History: Essays in Interpretation* (Philadelphia, 1966), pp. 349-50.

88. Only 957 harvest excursionists came from the United States in 1917, and 224 in 1918, compared to 5,000 in 1911 and 2,204 in 1926. Thompson, " 'Bringing in the Sheaves,' " p. 472.

89. NAC, RG76, Vol. 619, File 917093, Part 1, J.B. Walker to R.J. Reid, 28/8/1917.

90. "Threshing in Canada," *Industrial Solidarity*, 21/10/1916.

91. *Albertan*, 22/8/1917. I am indebted to Paul Voisey for bringing this information to my attention.

92. NAC, RG76, Vol. 619, File 917093, Part 1, J.B. Walker to R.J. Reid, 28/8/1917; S.R. Waugh, RNWMP *Report*, 3/8/1917.

93. *Ibid.*, J.B. Walker to W.D. Scott, 11/8/1917.

94. *Ibid.*, 18/9/1917.

95. *Ibid.*, 2/8/1917.

96. *Ibid.*, "I.W.W. Not Allowed to Cross into Canada," *Montana Record-Herald*, 14/9/1917.

97. *Ibid.*, J.E. Brook to W.D. Scott, 28/8/1917.

98. *Ibid.*, S.R. Waugh, RNWMP *Report*, 3/8/1917.

99. *Ibid.*, W.D. Scott to Dunlop, 23/7/1917.

100. On the Lintz case, see also Barbara Roberts, *Whence They Came: Deportation from Canada, 1900-1935* (Ottawa, 1988), pp. 76-77.
101. "Threshing in Canada," *Industrial Solidarity*, 21/10/1916.
102. PC 2384, 28/9/1918.
103. SAB, M12/II/39, Motherwell Papers: European War: Farm Labour Exemptions, p. 5608, "The Report of the Special Committee on Farm Labour."
104. *Ibid.*, pp. 5608, 5609.
105. See, for example, PAM, MG14/B45, Valentine Winkler Papers; SAB, M6/X-203-3, Charles Dunning Papers, Exemptions from Military.
106. SAB, M6/X-203-3, Charles Dunning Papers, Exemptions from Military, p. 42622, Isaac West to Charles Dunning, 15/5/1918.
107. Author's calculations. *Canada Yearbook*, 1914, p. 181; see Appendix, Table 1, Agricultural Work Force.
108. Canada, Dominion Experimental Farms, *Labour Saving on the Farm*, Special Circular No. 16 (Ottawa, 1918), pp. 1-2.
109. PAM, MG10/E1, United Farmers of Manitoba Papers, Box 15, *Manitoba Grain Growers' Year Book, 1918*, "Report of Proceedings of Annual Convention," Brandon, 1918.
110. Thompson, *Harvests of War*, p. 154.
111. Dominion Experimental Farms, *Labour Saving on the Farm*, pp. 1-2.

Chapter 7

1. PAA, 80.218/20, Thomas E. Stillwell interview, 17/9/1978.
2. NAC, MG30, Vol. C87, John Grossman Papers, ms., "Why I Left Germany," pp. 12-14.
3. Canada, Royal Grain Inquiry Commission, *Report of the Royal Grain Inquiry Commission*, p. 23.
4. Fowke, *The National Policy and the Wheat Economy*, pp. 78-80; G.A. Elliott, "Problems of a Retrograde Area in Alberta," in W.A. Mackintosh, *Economic Problems of the Prairie Provinces* (Toronto, 1935), pp. 291-94.
5. GAI, BN .C212 G, CPR Papers, P.L. Naismith Correspondence, File 154, P.L. Naismith to Sir Augustus Nanton, 4/1/1922.
6. Murchie, *Agricultural Progress on the Prairie Frontier*, p. 20. See also Jones, *Empire of Dust*.
7. Murchie, *Agricultural Progress on the Prairie Frontier*, p. 20.
8. See Appendix, Table 4, Prairie Farms.
9. Fowke, *The National Policy and the Wheat Economy*, p. 78.
10. Vernon C. Fowke, "The Myth of the Self-Sufficient Canadian Pioneer," *Transactions of the Royal Society of Canada*, LIV, Series III (June, 1962); Gavin Wright, "American Agriculture and the Labour Market: What Happened to Proletarianization?" *Agricultural History*, 62, 3 (1988).
11. See, for example, E.C. Morgan, "Soldier Settlement in the Prairie Provinces," *Saskatchewan History* (Spring, 1968); R.A. MacDonell,

"British Immigration Schemes in Alberta," *Alberta Historical Review* (Spring, 1968).

12. Martin, *'Dominion Lands' Policy*, p. 172.

13. Calculated from figures in Appendix, Table 3-B, Homestead Cancellations.

14. Calculated from figures in Appendix, Table 3-A, Homestead Entries.

15. W. Kaye Lamb, *History of the Canadian Pacific Railway* (New York, 1977), p. 255.

16. Canadian National Railways, *Own a 'Selected Farm' to fit your needs along the line of the Canadian National Railways. Read, think, act* (Chicago, 1925), p. 3; Canadian Pacific Railway, *Tell Me – why should I leave my farm and home and move to Western Canada?* (Winnipeg, 1924), p. 3.

17. Saskatchewan, Royal Commission of Inquiry into Grain Markets, *Report of the Agricultural Credit Commission of the Province of Saskatchewan* (Regina, 1913), p. 20; E.C. Hope, "What the Price of Wheat Means to Western Farmers," *Proceedings of the Conference on Markets for Western Farm Products* (Winnipeg, 1938), p. 149.

18. Hedges, *Building the Canadian West*, pp. 315-16.

19. Lamb, *History of the Canadian Pacific Railway*, p. 339.

20. "Equipment for a Half-Section Farm," *Nor'-West Farmer*, 20/2/1920, p. 230.

21. See Appendix, Table 1-A, Annual Farm Workers' Wages.

22. PAM, MG 10/E1, United Farmers of Manitoba Papers, pp. 924-25, L.E. Matthaei to J.W. Ward, 8/1/1926, and p. 922, John Ward to L.E. Matthaei, 8/2/1926.

23. See Chapter 5, note 6.

24. Danysk, "'Showing These Slaves Their Class Position,'" pp. 167-68.

25. TFL, ms. 179, Robert S. Kenny Collection, Box 9, J.M. Clarke, "Draft Agrarian Program of the CPC," June, 1930. For a discussion of Clarke's theorization on labour-capital relations in prairie agriculture, see David Monod, "The Agrarian Struggle: Rural Communism in Alberta and Saskatchewan, 1926-1935," *Histoire sociale/Social History*, XVIII, 35 (May, 1985), pp. 103-08.

26. For material on farm labour organizing in the United States, see [Stuart Jamieson], *Labor Unionism in American Agriculture*, United States Department of Labor Bulletin 836 (Washington, 1945); Melvyn Dubofsky, *We Shall Be All: A History of the Industrial Workers of the World* (Chicago, 1969).

27. See Appendix, Table 4, Prairie Farms; Table 1, Agricultural Work Force. Author's calculations.

28. The first census year for which figures are available is 1936. Of the 26,475 permanent male employees reported on farms, 19,738 or 74.6 per cent were the only full-time hands on each farm. Dominion Bureau of

Statistics, *Census of the Prairie Provinces*, 1936, Vol. ii, Table 144, pp. 321, 795, 1243; Table 208, pp. 353, 827, 1275. In a survey of thirteen farming communities during the 1930s, George Britnell found that hired labour never accounted for more than "one-half an adult unit in the farm household." George E. Britnell, *The Wheat Economy* (Toronto, 1939), p. 162.

29. "Life in the Harvest Fields," *OBU Bulletin*, 16/8/1928, p. 1.

30. TFL, Kenny Collection, ms. 179/9, J.M. Clarke, "Memo on the Agricultural Proletariat," 26/4/1930, p. 3.

31. NAC, RG76, Vol. 619, File 917093, Part 1, S.R. Waugh, *RNWMP Report*, 3/8/1917.

32. GAI, BN.C212G/753, CPR Papers, James Colley Correspondence, memo from J. Colley, 14/9/1928.

33. See PC 183 for immigration guidelines during the 1920s. Central and eastern Europeans came under sections 1, 2, 3, 4, and 8. The Railway Agreement came into effect on September 5, 1925, and was renewed in 1928. England, *The Colonization of Western Canada*, pp. 82-84; Donald Avery, *"Dangerous Foreigners" : European Immigrant Workers and Labour Radicalism in Canada, 1896-1932* (Toronto, 1979), pp. 100-01.

34. See, for example, Varpu Lindstrom-Best, "'I Won't Be a Slave!' – Finnish Domestics in Canada, 1911-30," in Jean Burnet, ed., *Looking into My Sister's Eyes: An Exploration in Women's History* (Toronto, 1986); Jorgen Dahlie and Tissa Fernando, eds., *Ethnicity, Power and Politics in Canada* (Toronto, 1981); Ian Radforth, *Bushworkers and Bosses: Logging in Northern Ontario, 1900-1980* (Toronto, 1987); Donald Avery, "Continental European Immigrant Workers in Canada, 1896-1919: From 'Stalwart Peasants' to Radical Proletariat," *Canadian Review of Sociology and Anthropology*, 12, 1 (1975).

35. C.H. Young, *The Ukrainian Canadians* (Toronto, 1931), p. 52.

36. PAA, 73.307/54, Department of Agriculture Papers: General Administration, memorandum for Mr. Stewart, 18/6/1927.

37. GAI, BN. C212 G, CPR Papers, James Colley Papers, Farm Help Applications.

38. Avery, *"Dangerous Foreigners"*; Ruth Bleasdale, "Class Conflict on the Canals of Upper Canada in the 1840's," *Labour/Le Travailleur*, 7 (1981).

39. SAB, RC-M6/13, Saskatchewan Royal Commission on Immigration and Settlement, 1930, Vol. vii:13, Sebastian Holizki.

40. NAC, RG76/234/135755, Part 4, Assisted Passage British Workers, c. 1925.

41. PAA, 65.118, Employment Service Papers, F.H. Dudley to W. Smitten, 5/2/1930.

42. "Revised Statutes of Manitoba, 1913," Masters and Servants Act, R.S., c.108, s.1, ch. 124; "Revised Statutes of Saskatchewan, 1920," Masters

and Servants Act 1918-19, c.61, s.2, ch. 205; "Revised Statutes of Alberta, 1922," Masters and Servants Act 1904, c.3, s.7, ch. 180, all in Canada, Department of Labour, *Labour Legislation in Canada as existing December 31, 1928* (Ottawa, 1929), pp. 392-94 (Manitoba), pp. 486-87 (Saskatchewan), pp. 531-32 (Alberta).

43. SAB, M6/X-11-0, Charles Dunning Papers, "Memo on Farm Wages," c. 1922.

44. *Ibid.* C.A. Dunning to Ira B. Cushing, 16/8/1921.

45. Canada, Department of Immigration and Colonization, *Assisted Settlement of Approved British Families on Canadian Government Farms*, 1925, p. 4.

46. "Go to a Farm or Go to Jail," *OBU Bulletin*, 15/12/1927.

47. GAI, BN.C212 G, CPR Papers, File 678, B.A. Pickel to J. Colley, 13/11/1926 and 9/12/1926.

48. Saskatchewan, Department of Agriculture, *Practical Pointers for Farm Hands* (Regina, 1915), *passim*.

49. PAA, 70.414/407, Alberta *Sessional Papers*, W. Frantzen to Director of Employment Service, 4/9/1920.

50. PAA, 72.370/6a, Attorney General Papers, Alberta Provincial Police Report, "C" Division, W. Brankley, 1923.

51. *Ibid.*, T. Hidson, 21/10/1925.

52. PC 2384, 28/9/1918 to 2/4/1919.

53. Roberts, *Whence They Came*, pp. 71-97.

54. "What the I.W.W. Offers to the Farmer and the Farm Hand," *The One Big Union Monthly*, II, 12 (Dec., 1920), p. 34.

55. "Harvesters' Strike is on in Canada," *Industrial Solidarity*, 8/10/1921.

56. NAC, RG 76, Vol. 619, File 917093, Part 2, R.S. Knight to The Commissioner, RCMP, Ottawa, 10/9/1921.

57. *Ibid.*, Thomas Gelley to The Secretary, Department of Immigration, 2/11/1922.

58. *Ibid.*, Part 1, Bruce Walker to W.D. Scott, 18/9/1917.

59. *Ibid.*, Part 2, Thomas Gelley to The Secretary, Department of Immigration, 2/11/1922.

60. *Ibid.*, "Deportation Proceedings: Samuel Scarlett," p. 7. See also Donald Avery, "British-born 'Radicals' in North America, 1900-1941: The Case of Sam Scarlett," *Canadian Ethnic Studies*, X, 2 (1978); Roberts, *Whence They Came*, pp. 92-94.

61. NAC, RG 76, Vol. 619, File 917093, Part 2, A.L. Jolliffe to T. Gelley, 6/8/1923.

62. *Ibid.*, C.B. Fryett to F.W. Schultz, 5/9/1923.

63. SAB, AtG/4/C, Attorney General Papers, Saskatchewan Provincial Police, *Annual Report*, Weyburn, 1923.

64. NAC, RG 76, Vol. 619, File 917093, Part 2, R.F.V. Smyly to F.W. Schultz, 3/9/1923.

65. Canada, Department of Labour, "I.W.W. Reappears in Canada," *Report on Labour Organization in Canada*, 1923, p. 182.

66. "Canadian Harvest is Paying Better Wages," *Industrial Solidarity*, 29/9/1923.

67. Canada, Department of Labour, *Report on Labour Organization in Canada*, 1924, p. 173.

68. *Ibid.*, 1925, pp. 176-77.

69. SAB, AtG/4/C/7, Attorney General Papers, SPP *Annual Report* 1923; PAA, 72.370/7a, Attorney General Papers, APP *Annual Report*, 1924; SAB, M6/Y, Charles Dunning Papers, Charles Dunning to W.J. DeGrow, 12/9/1922.

70. WRA, 130/44/3, IWW Papers, *Minutes of the 16th Constitutional General Convention of the IWW*, 13 October to 10 November 1924, p. 89.

71. *Report on Labour Organization in Canada*, 1924, p. 1173.

72. PAA, 72.370/7a, Attorney General Papers, APP *Annual Report*, 1924.

73. Norman Penner, *The Canadian Left: A Critical Analysis* (Toronto, 1977).

74. PAA, 72.370/7a, Attorney General Papers, APP *Annual Report*, 1924.

75. PAA, 75.126/4613, Attorney General Papers, APP Report, F. Lesley, 13/8/1923.

76. WRA, 171, E.W. Latchem Collection, "Published Letter from Tom Doyle, Secy and Tom Connors, Chairman GOC AWIU 110 IWW to the Striking Railroad Shop Crafts," 28/7/1922. Emphasis in original.

77. "Farm Workers," *Searchlight*, 3/9/1920.

78. David Jay Bercuson, *Fools and Wise Men: The Rise and Fall of the One Big Union* (Toronto, 1978), p. 166.

79. "Agricultural Workers Organizing in O.B.U.," *The Searchlight*, 28/5/1920, p. 3.

80. PAA, 72.159/23, Farmilio Papers, *Minutes of General Convention, OBU, January 1921*, "General Executive Report," 56 B. p. 1.

81. PAM, MG 10/A3, One Big Union Papers, "A report from Midgley, General Executive Board to All Units of the OBU," Winnipeg, 2/2/1921; Bercuson, *Fools and Wise Men*, p. 167.

82. "Agricultural Workers Organizing in O.B.U.," *The Searchlight*, 28/5/1920, p. 3.

83. PAM, MG 10/A 14-2, R.B. Russell Papers, #18, *Minutes of the OBU General Executive Board*, 27/2/1924.

84. *Ibid.*, *Minutes of Meeting of the Resident Members of the G.E.B. of the O.B.U.*, 5/5/1924.

85. Bercuson, *Fools and Wise Men*, p. 230.

86. PAM, MG 10/A 14-2, R.B. Russell Papers, # 19, *Minutes of General Executive Meeting*, 29/7/1924; *Minutes of the Joint Executive Board*, 30/9/1924.

87. *Ibid.*, # 18, *Minutes of Joint Executive Boards of the G.E.B and the C.L.C. of the O.B.U.*, 14/4/1925.

88. "In the Harvest Fields," *OBU Bulletin*, 24/9/1925.

89. PAM, MG 10/A 14-2, R.B. Russell Papers, # 24, *Minutes of Joint Executive Board of the O.B.U.*, 26/6/1928.

90. "Life in the Harvest Fields," *OBU Bulletin*, 16/8/1928.

91. *Ibid.*

92. William Irvine, *The Farmers in Politics*, ed. and intro. by Reginald Whitaker (Toronto, 1976; originally published 1920); David Laycock, *Populism and Democratic Thought in the Canadian Prairies, 1910 to 1945* (Toronto, 1990).

93. Whitaker, "Introduction," in Irvine, *The Farmers in Politics*, p. xii.

94. W. Peterson in *Farm and Ranch Review*, 6/9/1920.

95. "The Sympathetic Strike," *Grain Growers' Guide*, 21/5/1919.

96. David Yeo, "Rural Manitoba Views the 1919 Winnipeg General Strike," *Prairie Forum*, 14, 1 (Spring, 1989). Yeo presents convincing evidence to challenge the widely accepted view that farmers were uniformly hostile to the strike. For the earlier interpretation, see, for example, W.L. Morton, *The Progressive Party in Canada* (Toronto, 1971; originally published 1950), pp. 117, 274; Martin Robin, *Radical Politics and Canadian Labour, 1880-1930* (Kingston, Queen's University, 1968), p. 206.

97. Morton, *The Progressive Party in Canada*, p. 227.

98. First minute book of Farmers' Union of Canada, Ituna Lodge No. 1, cited in Morton, *The Progressive Party in Canada*, p. 276.

99. Labour candidates ran in Moose Jaw (and won) and Regina (and lost) in 1921. The Moose Jaw candidate, William George Baker, had run under the labour banner since 1917. In the 1925 election he was re-elected as a Labor-Liberal candidate, but he was narrowly defeated in 1929 as a Liberal-Labor candidate. *Provincial Elections of Saskatchewan, 1905-1983*, 2nd edition (Regina, 1983), pp. 39, 45, 51.

100. Alvin Finkel, "The Rise and Fall of the Labour Party in Alberta, 1917-42," *Labour/Le Travail*, 16 (Fall, 1985).

101. *Alberta Labour News*, 4/9/1920, 2/4/1921.

102. *Edmonton Journal*, 9/7/1921, cited in Carl F. Betke, "Farm Politics in an Urban Age: The Decline of the United Farmers of Alberta After 1921," in Lewis H. Thomas, ed., *Essays on Western History* (Edmonton, 1976), p. 178.

103. See, for example, *Alberta Labour News*, 4/8/1923, 3/10/1926.

104. "Not the Whole Story," *Alberta Labour News*, 18/6/1927.

105. "Immigrants are Taking Laborers' Jobs," *Alberta Labour News*, 21/5/1927.

106. "United Farmers of Alberta," *Grain Growers' Guide*, 16/3/1921.

107. See Appendix, Table 2-A, Annual Farm Workers' Wages.

108. "The Labor Problem," UFA *Official Circular* #1, Calgary, 11/2/1921.

109. Manitoba Department of Agriculture, *Annual Report*, 1921, p. 61.

110. PAA, 75.490/1, Alberta Federation of Labour Papers, "Resolution 50," *Proceedings of the Eleventh Convention of the AFL*, November, 1924.

111. *Ibid.*, *Proceedings of the Fourteenth Convention of the AFL*, January, 1926.

112. *Ibid.*, *Proceedings of the Eighteenth Convention of the AFL*, January, 1930.

113. John Manley, "Communism and the Canadian Working Class During the Great Depression – The Workers' Unity League – 1930-1936" (Ph.D. thesis, Dalhousie University, 1984), p. 37.

114. PAA, 75.490/1, Alberta Federation of Labour Papers, *Proceedings of the Fifteenth Convention of the AFL*, January, 1928, p. 25.

115. *Ibid.*

116. PAA, 64.11/125, Legislative Assembly Papers, Report of Investigation Regarding Workmen's Compensation, 1918, Submission by H.W. Wood, 19/11/1917, pp. 1184, 1185, 1186.

117. Alberta *Statutes*, 1908, Workmen's Compensation Act: "This Act shall not apply to the employment of agriculture nor to any work performed or machinery used on or about a farm or homestead for farm purposes or for the purposes of improving such farm or homestead." Canada, Department of Labour, *Labour Legislation in Canada . . . 1915* (Ottawa, 1918), pp. 544-45. The Alberta *Revised Statutes*, 1922, emphasized the provisions in "Exemptions of Farm Labor": *Labour Legislation in Canada as existing December 31, 1928* (Ottawa, 1929), p. 515. In both Manitoba and Saskatchewan, the Act applied only to "railway, factory, mine, quarry or engineering work or building repair, construction or demolition." Manitoba Workmen's Compensation Act 1920, C.159, s.1; Manitoba Consolidated Amendments, 1924, ch. 209; Saskatchewan Workmen's Compensation Act, 1910-11, c.9, s.1; Revised Statutes of Saskatchewan, 1920, in *Labour Legislation in Canada 1928*, p. 439 (Manitoba), p. 496 (Saskatchewan).

118. TFL, Kenny Collection, Box 9, J.M. Clarke, "Memo on the Agrarian Question for Comrade Morris," March, 1930.

119. "Employer and Hired Man," *Employer and Farmers' Manual: Legal Advisor and Veterinary Guide* (Winnipeg, 1920), p. 152.

120. SAB, M4/I/100, W.M. Martin Papers, pp. 30017-18, W.M. Martin to G.D. Robertson, 1/8/1921.

121. PAM, MG 10/E1, United Farmers of Manitoba Papers, pp. 924-25, L.E. Matthaei to J.W. Ward, 8/1/1926; p. 922, John Ward to L.E. Matthaei, 8/2/1926.

122. Winnifred Reeve, "Alberta, 1923," *Alberta History*, 28, 1 (Winter, 1980), p. 30.

Chapter 8

1. PAA, 73.167, taped interview with Jens Skinberg by Ellen Nygaard and Reevan Dolgoy, 31/1/1973.

2. *Ibid.*

3. For a discussion of the intersection of domestic relations with productive relations, see Hedley, "Relations of Production of the 'Family Farm': Canadian Prairies," p. 73.

4. MacPherson and Thompson, "The Business of Agriculture," p. 260.

5. PAA, 69.289, Premier's Papers, File 503, Agricultural Labour, 1921-29, Employment Bureau, A. Redshaw to W. Smitten, 23/5/1928.

6. For a discussion of the persistence of traditional rural "yeoman" values and attitudes into the era of capitalist agriculture, see Allan Kulikoff, "The Transition to Capitalism in Rural America," *William and Mary Quarterly*, 3rd Series, XLVI (January, 1989), pp. 141-44. Kulikoff finds that "Yeomen were embedded in capitalist world markets and yet alienated from capitalist social and economic relations." In the prairie West, this applies to a much greater extent in theory than in practice, as evidenced by the disjuncture between ideology and reality. Conflicting attitudes were often held at the same time. Cf. Jeffrey Taylor, *Fashioning Farmers: Ideology, Agricultural Knowledge and the Manitoba Farm Movement, 1890-1925* (Regina, 1994).

7. Author's calculations. *Census of Canada*, 1921, Vol. II, Table 28. pp. 128-29, 134-37.

8. Author's calculations. The number of farm workers listed as heads of families was 12,669. *Census of Canada*, 1931, Vol. V, Table 45, pp. 786, 796, 802.

9. Author's calculations. The figures are 150,038 single and 23,632 married farm workers and farmers' sons, and 51,549 single and 195,822 married farmers. *Ibid.*, Vol. VII, Table 54, pp. 644-45.

10. NAC, RG76, Vol. 234, File 135755, Part 5, John Barnett to W.J. Egan, 2/4/1927. See other letters in this file. See also GAI, BN.C212 G, CPR Papers, James Colley Correspondence, P.L. Naismith Correspondence.

11. *NAC*, RG76, Vol. 234, File 135755, Part 6, C.W. Vernon to Barnett, 20/5/1927.

12. For example, the CPR's Colonization Department, which was involved in farm labour placement, reported in 1926 that the local colonization boards placed orders for only fifty-three married couples and ninety-eight families, but 13,494 single men. This was typical of placements throughout the decade. GAI, BN.C212 G, CPR Papers, Advisory Committee Papers, "The Departmental Organization in Western Canada," [1926]. See also Danysk, "Farm Apprentice to Agricultural Proletarian," p. 30.

13. See, for example, NAC RG76, Vol. 234, File 13577, Part 5, John Barnett to W.J. Egan, 2/4/1927, and other letters in the file.

14. Manitoba Department of Agriculture, *Annual Report*, 1920, p. 12.

15. "What your Ambitious Friends are doing," *The Farm and Ranch Review*, 5/11/1920, p. 6.

16. Duncan Marshall, *Farm Management* (n.p.: Imperial Oil Ltd., 1931), p. 65.

17. Maurice Fitzgerald, "The Status of Farm Labour in Saskatchewan" (M.A. thesis, McMaster University, 1926), p. 35. This thesis contains much personal observation and is useful as a primary source.

18. *Census of Canada*, 1921, Vol. IV, Table 4, pp. 242-45, 270-71, 292-93; *ibid.*, 1931, Vol. VII, Table 40, pp. 134-35, 146-47, 156-57. Author's calculations.

19. "Wants Farmer Protected," *Nor'-West Farmer*, 20/2/1920, p. 230.

20. "Labor for Live Stock," *Nor'-West Farmer*, 20/7/1920, p. 1057.

21. GAI, BN .C212 G, CPR Papers, P.L. Naismith Correspondence, File 189, P.L. Naismith to H.A. Kuhn, 25/2/1925.

22. See, for example, Mrs. George [Marian] Cran, *A Woman in Canada* (Toronto, n.d. [c. 1908]), pp. 108-10; Susan Jackell, ed., *A Flannel Shirt and Liberty: British Emigrant Gentlewomen in the Canadian West, 1880-1914* (Vancouver, 1982).

23. Mark Rosenfeld has found that even in a non-frontier area, men who were temporarily without their wives engaged in a "bachelor culture of recreation." Mark Rosenfeld, "'It was a hard life': Class and Gender in the Work and Family Rhythms of a Railway Town 1920-1950," Canadian Historical Association, *Historical Papers/Communications historiques* (Windsor, 1988), pp. 262-63.

24. Compare with similar attitudes toward other groups of single men: Robert Harney, "Men Without Women: Italian Migrants in Canada, 1885-1930," in Betty Boyd Caroli, Robert F. Harney, and Lydia F. Tomasi, eds., *The Italian Immigrant Woman in North America* (Toronto, 1978); Karen Dubinsky, *Improper Advances: Rape and Heterosexual Conflict in Ontario, 1880-1929* (Chicago, 1993), pp. 143-62.

25. "Is the Average Home Sanitary?" *Grain Growers' Guide*, 13/9/1922, p. 13.

26. *Grain Growers' Guide*, 23/9/1925, p. 18.

27. "A Mother of Two" to editor of "Sunshine" column, *Grain Growers' Guide*, 16/10/1912.

28. See, for example, John MacDougall, *Rural Life in Canada, Its Trends and Tasks* (Toronto, 1973), originally published 1913.

29. *Farmer's Advocate*, 16/6/1920, p. 1028, 17/11/1920, p. 1854.

30. Marjorie Harrison, *Go West – Go Wise! A Canadian Revelation* (London, 1930), p. 131.

31. David C. Jones, "'There is Some Power About the Land' – The Western Agrarian Press and Country Life Ideology," *Journal of Canadian Studies*, 17, 3 (Fall, 1982), p. 96.

32. *Grain Growers' Guide*, 1/4/1926, p. 31.

33. *Canadian Power Farmer*, June, 1923, p. 4.

34. "Do You Want Your Daughter to Marry a Farmer?" *Grain Growers' Guide*, 8/3/1922, p. 15. The prize-winning letters appeared in the June and July issues.

35. *Ibid.*

36. Fitzgerald, *Status of Farm Labour*, p. 81.

37. PAA, 69.289, Premier's Papers, Employment Bureau, File 508, F.W. Crandall to Greenfield, 5/7/1923.

38. GAI, BN .C212 G, CPR Papers, James Colley Correspondence, memorandum, James Colley to Vanscoy, 10/10/1928.

39. *Labour Gazette*, 1929, p. 496.

40. Nancy Forestall has discovered a "homosocial bachelor culture" in a predominantly male mining community in northern Ontario, in which the ideology of the family was largely absent and emphasis was placed on the "rough" rather than the "respectable" elements of masculinity. Nancy Forestall, "The Rough and Respectable: Gender Construction in the Porcupine Mining Camp, 1909-1920," paper presented at the Canadian Historical Association annual meeting, Kingston, 1991.

41. PAM, MG10/E1, United Farmers of Manitoba Papers, Box 12, United Farm Women of Manitoba Survey of Farm Homes, Questionnaire, 1922.

42. Dean Davenport to the Industrial Commission, 1899, cited in Paul Taylor, "The American Hired Hand: His Rise and Decline," *Land Policy Review*, VI, 1 (Spring, 1943), p. 12.

43. Kathleen Strange, *With the West in Her Eyes: The Story of a Modern Pioneer* (Toronto, 1937), p. 253. See also Barons History Book Club, *Wheat Heart of the West* (Barons, Alberta, 1972), p. 246.

44. GAI, BN .C212 G, CPR Papers, James Colley Correspondence, File 692, James Colley to Vanscoy, 7/4/1926.

45. *Ibid.* File 1484, James Colley to Robert Stuart, 25/10/1925.

46. "Hired Men Want to Quit," *Nor'-West Farmer*, 20/7/1920, p. 1052.

47. Paul Willis, "Shop Floor Culture, Masculinity and the Wage Form," in Clarke, Critcher, and Johnson, eds., *Working Class Culture*, p. 192.

48. GAI, 630.3 A278, *Agricultural Alberta*, January, 1921, cover.

49. Advertisement for J.I. Case Threshing Machine, *Grain Growers' Guide*, 17/2/1926, p. 9. See also the regular *Guide* column, "What's New in Farm Implements." Other farm journals also kept their readers up-to-date about the latest technical developments and machinery. See especially the *Canadian Power Farmer*.

50. Braverman, *Labor and Monopoly Capital*; Paul Thompson, *The Nature of Work: An Introduction to Debates on the Labour Process* (London, 1983); Stephen Wood, ed., *The Degradation of Work? Skill, deskilling and the labour process* (London, 1982); Craig Littler, *The Development of the Labour Process in Capitalist Societies: A Comparative Study of the Transformation of Work Organization in Britain, Japan and the USA* (London,

1982). For studies that apply labour process theory, see the essays in Craig Heron and Robert Storey, eds., *On the Job: Confronting the Labour Process in Canada* (Kingston and Montreal, 1986); Joel Novek, "Grain Terminal Automation: A Case Study in the Control of Control," *Labour/Le Travail*, 22 (Fall, 1988).

51. "Bonanza Farming," *Grain Growers' Guide*, 15/9/1926, p. 7.

52. "The Hired Man," a paper read by G.C.D. Edwards at the Qu'Appelle Farmers' Institute, *Nor'-West Farmer*, 20/9/1899, p. 701.

53. PAA, 65.118, Employment Service Papers, *Employment Service Conference*, 1921, p. 6.

54. Charles More, *Skill and the English Working Class, 1870-1914* (London, 1980), p. 15. See also Littler, *Development of the Labour Process*, pp. 7-14.

55. Braverman, *Labor and Monopoly Capital*, p. 433.

56. Ken Kusterer, *Know-How on the Job: The Important Working Knowledge of "Unskilled" Workers* (Boulder, Colorado, 1978), esp. pp. 177-80.

57. More, *Skill and the English Working Class*, pp. 16-26; Littler, *Development of the Labour Process*, pp. 8-9; Radforth, *Bushworkers and Bosses*, pp. 67-69 and *passim*.

58. PAM, MG8/B65, Arthur Sherwood Papers, Sherwood to father, 22/7/1882.

59. GAI, BN .C212 G, CPR Papers, James Colley Correspondence, File 692, Geo. C. Peattie to J. Dennis, 1/10/1926.

60. PAA, 65.118/17/47, Employment Service Papers, W.D. Trego, Labor Committee, United Farmers of Alberta, *Official Circular #10*, 29/3/1921.

61. GAI, BN.C212 G, CPR Papers, James Colley Correspondence, File 692, Geo. C. Peattie to J. Dennis, 1/10/1926.

62. *Ibid.*, James Colley to Vanscoy, 7/4/1926.

63. NAC, MG 30 C63, Noel Copping Papers, "Prairie Wool and Some Mosquitoes," excerpts from a diary, Saskatchewan, 1909-1910, p. 19.

64. Maxwell, *Letters home*, p. 37.

65. Kusterer, *Know-How on the Job*, p. 179.

66. "Bonanza Farming," *Grain Growers' Guide*, 15/9/1926, p. 7.

67. Friedmann, "World Market, State, and Family Farm."

68. F.C. Birchall, "Man Power vs. Barn Machinery," *Nor'-West Farmer*, 20/2/1920, p. 234.

69. Mackintosh, *Economic Problems of the Prairie Provinces*, p. 16.

70. Andrew Stewart, "Trends in Farm Power and Their Influence on Agricultural Development," Appendix A in Murchie, *Agricultural Progress on the Prairie Frontier*; John Lier, "Farm Mechanization in Saskatchewan," *Tijdschrift Voor Economishce en Sociale Geografie*, LXII, 3 (Mei/Juni, 1971).

71. *Canadian Power Farmer*, 4/1/1920.

72. But for costs for horses and their equipment, see Robert Ankli, H. Dan Helsberg, and John Herd Thompson, "The Adoption of the Gasoline

Tractor in Western Canada," in Akenson, ed., *Canadian Papers in Rural History*, Vol. II, pp. 23-24.

73. Stewart, "Trends in Farm Power," p. 313.

74. *Ibid.*, p. 297.

75. See Appendix, Table 5-A, Tractor Sales; Table 5-C, Numbers of Horses, Alberta.

76. J.G. Taggart, "Tractor and Combine," in Duncan Marshall, *Field and Farm Yard* (n.p.: Imperial Oil Ltd., 1929), p. 201.

77. Ankli, Helsberg, and Thompson, "The Adoption of the Gasoline Tractor," p. 19.

78. See Appendix, Table 1-A, Annual Farm Workers' Wages.

79. Birchall, "Man Power vs. Barn Machinery," p. 235.

80. Advertisement for International Harvester Company of Canada, *Nor'-West Farmer*, 5/2/1921, p. 119.

81. Advertisement for Case Power Farming Machinery, *Farm and Ranch Review*, 5/7/1920, p. 15.

82. Advertisement for Litscher Lite Plant, *ibid.*, 20/7/1920, p. 8.

83. L.F.R.G., "A Tractor Advocate," *Nor'-West Farmer*, 21/2/1921.

84. See Appendix, Table 5-A, Tractor Sales.

85. Thomas Isern, "Adoption of the Combine on the Northern Plains," *South Dakota History*, 10 (Spring, 1960), pp. 105-11.

86. See Appendix, Table 5-B, Combine Sales.

87. L.A. Reynoldson, R.S. Kifer, J.H. Martin, and W.R. Humphries, *The Combined Harvester-Thresher in the Great Plains*, USDA Technical Bulletin 70 (Washington, 1928), p. 23, cited in Mary Wilma Hargreaves, *Dry Farming in the Northern Great Plains, 1900-1925* (Cambridge, Mass., 1957), p. 518n.

88. Taggart, "Tractor and Combine," p. 205.

89. Edna Tyson Parson, *Land I Can Own: A Biography of Anthony Tyson and the Pioneers Who Homesteaded with Him at Neidpath, Saskatchewan* (Ottawa, 1981), p. 71.

90. *Labour Gazette*, 1931, p. 866.

91. Koeppen in Knight, *Stump Ranch Chronicles*, p. 56.

92. PAA, 65.118/10, H.A. Craig to W. Smitten, 13/9/1929.

93. R. Bruce Shepherd, "Tractors and Combines in the Second Stage of Agricultural Mechanization on the Canadian Plains," *Prairie Forum*, 11, 2 (Fall, 1986), p. 265.

94. Giscard, *Dans La Prairie Canadienne*, p. 14.

95. MLL, Acc. 8 Jan 1951, Vertical File, "Local History – Manitou," W.N. Rolfe, "A Citizen of Canada," p. 8.

96. *Nor'-West Farmer*, 5/1/1920, p. 17.

97. GAI, BN.C212 G, CPR Papers, James Colley Correspondence, File 1484, A. Schoonover to James Colley.

98. Sims, *Elements of Rural Sociology*, p. 429.

99. Carl E. Solberg, *The Prairies and the Pampas: Agrarian Policy in Canada and Argentina, 1880-1930* (Stanford, California, 1987), p. 99; Argersinger and Argersinger, "The Machine Breakers"; Seager, "Captain Swing in Ontario?"

Chapter 9

1. Betty Kilgour, ed., *As the Years Go By* (Three Hills, Alberta, 1970), p. 189.
2. *Ibid.*, p. 159.

Index

Accidents, 208 n30

Agricultural Alberta, 155

Agricultural developments: pre-1900, 26-30; 1900-1918, 48-52; 1920s, 113-15

Agricultural ladder, 52-56, 59, 90, 118, 173; United States, 42, 53-56, 152; Ontario 53-54

Alberta, 16, 50, 80, 104, 106, 107, 114, 116, 128, 130, 136, 137, 138; Beaver River, 114; Benton, 80; Beverly, 131; Calgary, 53, 168; Champion, 108; Cold Lake, 114; Dalum, 142-43; Drumheller, 137; Edmonton, 128, 138; Fort Saskatchewan, 138; Lacombe, 53; Lavoy, 69; Lethbridge, 53, 79, 137; Medicine Hat, 137; Nanton, 71; Olds, 55, 90, 160; Peace River, 73, 114; Penhold (NWT,) 9; Raymond, 143; Stettler, 53, 82; Three Hills, 172; Vulcan, 107

Alberta Employment Service, 79, 137, 158

Alberta Federation of Labour, 136, 137-38

"Alberta Homesteader, The," 72

Alberta Labour News, 136

Albertan, The, 107

Alienation, 97, 175, 209 n51

Amey, Edward, 40, 41

Ankli, Robert, 34

Appleyard, Joshua, 32

Apprenticeship, 9, 52-63, 64, 67, 70, 88, 98, 99, 144, 151, 161, 175, 202 n40

Arch, Joseph, 39, 41

Armitage, J.S., 77

Bachelor's Paradise, A, 73-74

Balcarres Grain Growers' Cooperative Association, 103

Baldwin, Harry, 46-48, 99-100

Barneby, Henry, 33

Becker, George, 9, 10

Begg, Alexander, 32-33

Binnie-Clark, Georgina, 88, 100

Blair, W.J., 104

Booth, Si, 100

Boske, Hans, 72

Bracken, John, 49

Braverman, Harry, 156-57, 158

British Columbia, 24; Fernie, 107; Sandon, 133; Vancouver, 104, 128, 142

British Empire Service League, 123

Burgeson, J.B., 104

Canada Food Board, 106

Canadian Council of Agriculture, 118

Canadian Credit Mens' Trust Association, 211 n85

Canadian Labour Party, 136

Canadian National Railways, 122, 126

Craig Heron,
Working in Steel: The Early Years in Canada, 1883-1935, 1988.
ISBN 0-7710-4086-5

Wendy Mitchinson and Janice Dickin McGinnis, Editors,
Essays in the History of Canadian Medicine, 1988.
ISBN 0-7710-6063-7

Joan Sangster,
Dreams of Equality: Women on the Canadian Left, 1920-1950, 1989.
ISBN 0-7710-7946-X

Angus McLaren,
Our Own Master Race: Eugenics in Canada, 1885-1945, 1990.
ISBN 0-7710-5544-7

Bruno Ramirez,
On the Move:
French-Canadian and Italian Migrants in the
North Atlantic Economy, 1860-1914, 1991.
ISBN 0-7710-7283-X

Mariana Valverde,
"The Age of Light, Soap, and Water":
Moral Reform in English Canada, 1885-1925, 1991.
ISBN 0-7710-8689-X

Bettina Bradbury,
Working Families:
Age, Gender, and Daily Survival in Industrializing Montreal, 1993.
ISBN 0-7710-1622-0

Andrée Lévesque,
Making and Breaking the Rules: Women in Quebec, 1919-1939, 1994.
ISBN 0-7710-5283-9

Cecilia Danysk,
Hired Hands: Labour and the Development of Prairie Agriculture,
1880-1930, 1995.
ISBN 0-7710-2552-1

RELUCTANT HOST:

Canada's Response to Immigrant Workers, 1896-1994

Donald H. Avery

Reluctant Host is a concise historical analysis of the evolution of Canadian immigration policy this century with three major themes: how pressure groups – business, labour, ethnic, political, bureaucratic – determined immigration policies; a study of the experiences of immigrants as shaped by racial and ethnic considerations; and an assessment of official policy in relation to class, race, and ethnicity of immigrant workers. Canada's immigration experience is compared and contrasted with other major immigration nations, most notably the United States.

Based on extensive archival research, Avery explores the working and living conditions of immigrant workers as they adjusted to the Canadian environment. Although Avery addresses the experiences of professional and skilled migrants, his emphasis is on the massive numbers of unskilled immigrant workers needed for Canada's labour-intensive industries. The role of the Canadian state in trying to control the influx of certain groups of so-called "dangerous foreigners" from "non-preferred" countries is a central focus of the book.

Reluctant Host provides students of ethnic, labour, social, and Canadian history with a comprehensive analysis of Canadian immigration policy, and how it has evolved over the years.

ISBN 0-7710-0827-9 $24.95 paper   342 pages   8 pages b&w photos